Air Ions: Physical and Biological Aspects

Editors

Jonathan M. Charry, Ph.D.
Environmental Research Information, Inc.
New York, New York

Robert Kavet, Sc.D.
Health Effects Institute
Cambridge, Massachusetts

CRC Press, Inc.
Boca Raton, Florida

Library of Congress Cataloging-in-Publication Data
Air ions.

Bibliography: p.
Includes index.
1. air, Ionized--Physiological effect. I. Charry, Jonathan M. II. Kavet, Robert I.
QP82.2.I5A38 1987 574.19'1 86-29185
ISBN 0-8493-6535-X

This book represents information obtained from authentic and highly regarded sources. Reprinted material is quoted with permission, and sources are indicated. A wide variety of references are listed. Every reasonable effort has been made to give reliable data and information, but the author and the publisher cannot assume responsibility for the validity of all materials or for the consequences of their use.

Direct all inquiries to CRC Press, Inc., 2000 Corporate Blvd., N.W., Boca Raton, Florida, 33431.

International Standard Book Number 0-8493-6535-X

Library of Congress Card Number 86-29185
Printed in the United States

PREFACE

An air ion, for practical purposes, is a cluster of molecules in the atmosphere that possesses either a net positive or negative charge. Interest in the physical/chemical properties and biological effects of air ions dates back to the early 1900s. Perhaps one of the reasons for this long-standing interest is that air ions are formed continuously by various natural forces and are, therefore, ubiquitous. For example, ions are created by waterfalls, solar activity, radioactivity in soil and rock, and by rapidly moving air or wind.

More recently it has been recognized that certain human activities or technologies can alter the charge concentration in the air. For example, combustion and grinding processes generate ions; on the other hand, depleted ion concentrations tend to occur in synthetic indoor environments supplied with artificial ventilation. Very importantly, direct-current (DC) overhead transmission lines generate plumes of both positive and negative air ions in the line vicinity.

The introduction of the DC technology in the U.S. in the 1970s stimulated renewed interest in the assessment of the potential biological effects of exposure to altered concentrations of air ions. This recent research, although triggered by the interest in transmission lines, is general enough to help determine whether ions in other environmental settings or under other conditions may initiate biological responses. The new research, with its emphasis on good experimental design and state-of-the-art techniques for ion generation and measurement, will, in all probability, help to clarify much of the conflicting data that have been published previously.

Significant advances in understanding the physical behavior of ions, in developing engineering solutions to the use of direct current transmission systems, and in designing exposure systems for bioeffects research have been brought about by two major efforts in the U.S. in recent years. The Electric Power Research Institute has created a systematic approach to assessing the biological effects of air ions at levels and concentrations characteristic of transmission lines either currently in operation or planned for the future. The U.S. Department of Energy, particularly at the Bonneville Power Administration, has developed a research program in physics and engineering to characterize accurately the physical behavior of air ions in the vicinity of operating transmission lines, and, as a result, has contributed to substantially improved methods and instrumentation for characterizing air ions.

Air Ions: Physical and Biological Aspects fully develops two areas that are important for a comprehensive understanding of the subject of air ions: (1) the physical/chemical nature of ions, and (2) their potential interaction with biological systems. The reader is led through a series of nine chapters, the first five of which lay the basis for understanding ions in the context of naturally and artificially created environments. The final four chapters are well situated to discuss the literature and history connected with the search for ion-induced biological effects.

Chapter 1, by Dr. Bracken, provides a description of the physical and chemical properties of air ions, explaining how they are formed and discussing factors that influence their evolution and lifetime. Chapter 2, by Drs. Bracken and Johnson, describes various environmental settings in which ambient air ion concentrations are altered, with emphasis on the ion environment that exists in the immediate vicinity of overhead DC transmission lines. Chapter 3, by Dr. McKnight, comprehensively details the state-of-the-art measurement techniques applicable to quantitative descriptions of air ion environments. In Chapter 4, by Dr. Kaune, air ions are discussed within the context of transit-time analysis, which defines the exposure limits for any physical configuration, ranging from transmission lines to laboratory exposure systems. Chapter 5, by Dr. Weigel, discusses state-of-the-art ion exposure chambers for animal experimentation, detailing considerations necessary for proper design and construction. Chapter 6, by Dr. Charry, is a review of the biological and health literature concerned with the exposure of humans and animals to air ions. Chapter 7, by Dr. Bailey, critically assesses the literature in which exposure of animals to air ions is associated with subsequent changes in blood and tissue levels of serotonin. Chapter 8, by Dr. Kavet, explores the available evidence for biological reactions to air ions and considers a mechanism that might pausibly explain the data to date. Finally, in Chapter 9, by Dr. Wehner, a history of air ion research, from its earliest inception, is provided in order to illustrate the critical themes and major hypotheses that have played a role in the development of scientific research that examined for biological effects of air ions.

We trust this volume serves as a useful reference for all those interested in the subject of air ions.

Jonathan M. Charry
Robert Kavet

THE EDITORS

Jonathan M. Charry, Ph.D. is President and Director of Research Laboratories, Environmental Research Information, Inc. (ERI, Inc.), a research consulting firm headquartered in New York City. ERI, Inc. is involved with a broad variety of environmental health issues of concern to both private industry and government.

From 1983 to 1986 Dr. Charry held an appointment as Senior Research Scientist in the Laboratory of Neuropharmacology and Environmental Toxicology at the New York State Institute for Basic Research. He was an Assistant Professor at the Rockefeller University from 1978 to 1983 and was a Rockefeller Foundation Fellow at The Rockefeller University from 1977 to 1978. Dr. Charry received his Ph.D. from New York University in 1976, and his A.B. from Tufts University in 1970.

Dr. Charry is past President and Director of the American Institute of Medical Climatology, past Chairman of the Section on Air and Other Environmental Ion Technology of the American Society of Testing and Materials, a member of The Rockefeller University Chapter of Sigma Xi, the American Association for the Advancement of Science, the American Psychological Association, the Bioelectromagnetics Society, the New York Academy of Sciences, the Eastern Psychological Association, and the Environmental Design Research Association.

Robert Kavet, Sc.D., is Vice President and Director of Program Operations for Environmental Research Information, Inc. (ERI, Inc.), a research consulting firm headquartered in New York City. ERI, Inc. is involved with a broad variety of environmental health issues of concern to both private industry and government.

Prior to joining ERI, Inc., Dr. Kavet was a senior staff scientist with the Health Effects Institute (HEI) in Cambridge, Mass., joining HEI in 1984. There, Dr. Kavet participated in the management of HEI's extramural research program and was also involved with specific environmental issues related to mobile source emissions, including gasoline refueling vapors, manganese, methanol, and oxidant gases.

Dr. Kavet received his Doctor of Science Degree in Respiratory Physiology in 1977 and Master of Science in Environmental Health Sciences in 1972, both from Harvard University. In 1978, following a year of post-doctoral research at the Harvard School of Public Health, he joined the Electric Power Research Institute (EPRI) in Palo Alto, Calif., where he served for $5^1/_2$ years as a project manager responsible for EPRI's program addressing the health issues related to electric and magnetic field environments, particularly those in proximity to AC and DC transmission lines.

Dr. Kavet received his Bachelor of Science in 1966 and Master of Electrical Engineering degree in 1967, both from Cornell University. He then worked for RCA Corporations's Aerospace Division in Burlington, Mass., where his focus was on analog and digital circuit design.

Dr. Kavet's principal interests and publishing activities involve the evaluation of potential health risks associated with exposure to airborne substances and exposure to electromagnetic environments near power facilities and in indoor environments. He is a member of the American Thoracic Society, the Bioelectromagnetics Society, and The AAAS.

ADVISORY BOARD

CONTRIBUTORS

William H. Bailey, Ph.D.
Department of Pharmacology
Cornell University Medical College
New York, New York

T. Dan Bracken, Ph.D.
T. Dan Bracken, Inc.
Portland, Oregon

Jonathan M. Charry, Ph.D.
Environmental Research Information, Inc.
New York, New York

Gary B. Johnson, Ph.D.
High Voltage Transmission Research Center
General Electric
Lenox, Massachusetts

William T. Kaune, Ph.D.
National Bureau of Standards
Boulder, Colorado

Robert Kavet, Sc.D.
Health Effects Institute
Cambridge, Massachusetts

Ronald H. McKnight, Ph.D.
National Bureau of Standards
Gaithersburg, Maryland

Alfred P. Wehner, D.D.S., Sc.D., Cand. Med.
Biology and Chemistry Department
Battelle Pacific Northwest Laboratories
Richland, Washington

Richard J. Weigel, Ph.D.
Department of Biology
Battelle Pacific Northwest Laboratories
Richland, Washington

We dedicate this book to the
memory of Dr. Albert Paul Krueger,
who pioneered the modern era of
biological studies of air ions and
was, above all, a true gentleman.

TABLE OF CONTENTS

Chapter 1

SMALL AIR ION PROPERTIES

T. Dan Bracken

TABLE OF CONTENTS

I. INTRODUCTION

The purpose of this chapter is to provide a general description of the physical and chemical properties of small air ions. An ion is an atom, molecule or cluster, or particle that has lost or gained one or more electrons and has thereby acquired a net electric charge. The term small ion excludes charged particles. In the context of this chapter, a small air ion is a cluster of molecules bound together by the Coulomb force associated with an excess or deficiency of electrons; whereas, a particle maintains its physical integrity regardless of its state of charge. Small air ions are the primary emphasis of this section. The presence of ions in the atmosphere was noticed by Coulomb in 1785.[1a] Since then, electrical charge in air and other electrical properties of the atmosphere have been investigated almost continuously. A thorough review of the ion literature related to atmospheric electricity is given in the classic volume by Chalmers.[1b] Research by atmospheric scientists on air ions has continued up to the present. In addition, interest in and research on air ions has been expanded to include such areas as high voltage transmission lines and particulate removal from high technology clean rooms. As indicated in other chapters, there has been considerable effort directed towards investigating possible biological effects of small air ions.

The presence of small ions is quantified by the ion density expressed in ions per unit volume. Historically, the unit, ions per cubic centimeter, has been used. The electrical charges residing on ions constitute a charge per unit volume or space charge density, expressed in units of Coulombs per cubic meter, C/m^3. The net space charge density is a measurable quantity and represents the free unbalanced charge in a volume taking no account of the charges of both signs which balance each other. Monopolar ion density and monopolar space charge density refer to the ion density and space charge density of one polarity, respectively. Usually, small ion densities are derived from measured space charge density by assuming one elementary charge per ion. This is a reasonable assumption because the small size of the ions, typically less than 2×10^{-9} m in diameter, makes it impossible for them to carry more than a single charge.

Atmospheres which contain air ions can be characterized as bipolar or unipolar. In a bipolar atmosphere, ions of both polarities are present in approximately equal numbers. The net electrical charge in such an atmosphere is relatively small and electric fields due to the space charge are also relatively small. Examples of bipolar ion environments are naturally occurring outdoor settings with concentrations of approximately 1000 ions/cm^3 for both polarities and certain clean room environments with 100,000 ions/cm^3 of both polarities. Unipolar atmospheres result from a biased source of ions that produces ions of a single polarity. The resulting electrical environment can be characterized as having a high net space charge density and electric fields due to this space charge. Examples of possible unipolar ion environments are the high voltage direct current (HVDC) transmission line environment and systems designed for exposure of animals to unipolar air ions.

In both types of ion environments, charged molecular clusters are extremely dilute. For example, a small ion density of 1000 ions/cm^3 represents less than 1 molecule out of 10^{17} which is charged.

II. SMALL ION GENERATION

A. Sources

Ionization refers to the process or the result of a process whereby a neutral atom or molecule acquires either a positive or negative charge. Ionization occurs when energy in excess of the electron ionization potential is absorbed by an atom yielding an electron and a positive ion. As discussed below these charged particles quickly evolve to more stable entities. The principal sources of ionization in outdoor air are cosmic radiation and alpha, beta, and gamma radiation from terrestrial sources. Other sources of ionization include UV radiation, sunlight, frictional charging due to wind, water droplet breakup, and natural electrical discharges such as lightning.

Production of ions by cosmic radiation is lowest near the equator and increases with latitude up to 50°. Production rates above this latitude do not change significantly.[2] At sea level the ion production rate at latitude 45° is about two ion pairs/cm^3/sec.[3] Ion production due to cosmic radiation increases with altitude. At an altitude of 2 km above sea level, the production rate is approximately three ion pairs/cm^3/sec.[3] Over the ocean cosmic radiation is the principal source of air ionization.

The radioactive decay of material in the earth and its atmosphere produces ionizing radiation. Because the concentrations of naturally occurring radioactive elements vary with geographical location, ion production rates due to this source vary with location. Anderson reports levels of about 11 ion pairs/cm^3/sec over granite in Finland and the Alps; about 75 ion pairs/cm^3/sec over Swedish alum shale; and about 2 ion pairs/cm^3/sec over sedimentary rocks such as sandstone.[4] Chalmers cites an average level of about four ion pairs/cm^3/sec due to radioactive material in the earth.[1]

Radon and thoron in the atmosphere produce approximately five ion pairs/cm^3 near the surface of the earth.[1] As with sources in the earth, levels of these radioactive gases will vary with location. As an extreme example, Wilkening reports ion pair production of 2300 ion pairs/cm^3/sec in Carlsbad Caverns where high levels of radon gas are present.[5]

Typical ion production rates near ground level are thus comprised of three sources:

Cosmic radiation	~ 2 ion pairs/cm^3/sec
Radioactive material in earth	~ 4 ion pairs/cm^3/sec
Radioactive material in air	~ 5 ion pairs/cm^3/sec
Total	10 to 12 ion pairs/cm^3/sec

Ions are also produced by natural localized sources such as lightning, waterfalls, and combustion where sufficient energy from electrical, mechanical, or chemical processes is available to cause ionization and separation of charge. Ions from these sources can be transported by electric fields or air currents and contribute to local ion levels. However, it is the production rate of 10 to 12 ion pairs/cm^3/sec due to naturally occurring ionizing radiation that generally determines the average outdoor ion level. Localized sources and their contributions to ion levels are discussed below.

Ionization also occurs in areas of strong electric field. In the case of energized electrodes such as conducting needles, wires, or power transmission line conductors, ionization near the electrodes results in a discharge in the air called corona. Corona occurs when a critical electric field (voltage gradient) is exceeded at the surface of the electrode. At this point free electrons near the electrode are accelerated in the field and achieve sufficient energy to ionize neutral molecules. Additional electrons are generated in each collision and these, in turn, cause additional ionization. Corona is a partial discharge, i.e., ionization occurs only in part of the interelectrode volume. Ions are also produced by a complete discharge such as an arc or lightning where ionization occurs along a path between electrodes. The corona phenomenon and, hence, the production of ions by this process, depend on many electrical and atmospheric parameters, including voltage gradient in the corona region, voltage waveform, composition of the electrodes, geometry of the electrodes, humidity, and barometric pressure. Electrical ionization processes and corona have been extensively investigated. The physics of ionization by corona discharge are discussed by Loeb.[6,7] Practical aspects of corona phenomena on transmission lines have also been widely studied.[8]

B. Evolution of Small Air Ions

During the ionization process, a free electron and positive atom or molecule are created.

The initial primary ions are N_2^+, O_2^+, N^+, and O^+. Free electrons attach within 10^{-8} to 10^{-7} sec of their liberation to neutral atoms and molecules forming negative ions primarily O_2^-. However, both positive and negative monomolecular ions are short-lived. Some of the positive and negative charges that are created combine with each other, thereby neutralizing one another in a process called recombination. At standard temperature and pressure, the molecular ions undergo approximately 5×10^9 collisions per second. During collisions with other constituents in the atmosphere, the primary monomolecular ions evolve into molecular clusters consisting of a kernel molecule with several surrounding water molecules attached to it.

The evolution of small air ions through their initial 10^{-3} to 10^{-2} sec of existence has been discussed by Huertas and Fontan[9] and Huertas et al.[10] for positive and negative ions, respectively. Their calculations involving more than 70 possible ion-molecule reactions for each polarity indicate that the major constituents of air, including water, equilibrate within milliseconds. After this time, further evolution depends on trace constituents of the atmosphere. Thus, the chemical composition of the hydrated ions continues to evolve as collisions with trace gases in the atmosphere occur. The actual composition of the fully evolved small ions will depend on the concentration and the electron or proton affinity of these trace constituents of the atmosphere.

Mohnen gives the following reaction times for trace constituents as a function of concentration at standard temperature and pressure:[3]

Concentration	Reaction time, sec
1 ppm; $1:10^{-6}$	0.4×10^{-3}
1 ppb; $1:10^{-9}$	0.4
1 ppt; $1:10^{-12}$	400

Since the typical lifetime of a small air ion in clean air is 10^2 to 10^3 sec, the small air ions have numerous collisions with any detectable impurity. Thus, trace gases at concentrations greater than 1 ppt can influence the nature of air ions.

Thus, the 10^2 to 10^3 sec lifetime of naturally generated small air ions allows interaction with essentially all constituents of the atmosphere. In contrast, the small air ions produced by corona or radioactive sources in experimental exposure systems often have lifetimes that are measured in seconds or less. In such cases small air ions will not have evolved to the same extent as long-lived ions. This is evidenced by higher mobilities measured for laboratory ions than for natural ions.[3] (Mobility is the velocity with which an ion travels in an electric field and is discussed below.) Eisele used this effect to demonstrate that the chemical composition of natural ions which he observed with a mass spectrometer were not generated by his instruments, i.e., short-lived ions generated at the inlet to his device had a simpler spectrum than natural ions.[11]

Small ions are removed from the atmosphere by three processes: recombination, attachment to larger particles, or contact with a surface. Attachment to aerosols is the principal method of removal in natural atmospheres.

C. Ion Equilibrium

The production and decay of small air ions in a volume can be described in terms of a differential equation with source terms and loss terms. The time rates of change of small positive and negative ion densities, n_1 and n_2, are given by

$$\frac{dn_1}{dt} = q - \alpha\, n_1 n_2 - \beta_{10} n_1 N_0 - \beta_{12} n_1 N_2 \qquad (1)$$

$$\frac{dn_2}{dt} = q - \alpha n_1 n_2 - \beta_{20} n_2 N_0 - \beta_{21} n_2 N_1 \tag{2}$$

where only neutral and singly charged aerosols have been assumed and q is the ion production rate in ions/cm³/sec, N is the density of aerosols, α is the recombination rate for small ions, and β_{ij} represents the combination coefficient for small ions of polarity i with aerosols of polarity j. Similar equations can be written for the large aerosol concentrations, N_i

$$\frac{dN_1}{dt} = Q_1 + \beta_{10} n_1 N_0 - \beta_{21} n_2 N_1 - \gamma N_1 N_2 \tag{3}$$

$$\frac{dN_2}{dt} = Q_2 + \beta_{20} n_2 N_0 - \beta_{12} n_1 N_2 - \gamma N_1 N_2 \tag{4}$$

where Q_i is the production rate for charged aerosols of polarity i, and γ is the aerosol-aerosol recombination coefficient. These equations have neglected transport of ions due to electric field and diffusion, which must be included to give a complete picture of ion dynamics. However, they still provide insight into bipolar atmospheres, as found in ambient conditions where electric fields are low.

Under natural conditions where ions are produced in pairs, the number of ions and charged aerosols of each polarity are approximately equal: $n_1 = n_2$, and $N_1 = N_2$. The generation rate of charged aerosols, Q_i, and the recombination rate of oppositely charged aerosols, γ, can also generally be neglected. In this case, equilibrium conditions, i.e., $\frac{dn}{dt} = \frac{dN}{dt} = 0$, result in the following aerosol fractions:

$$\frac{N_1}{N_0 + N_1 + N_2} = \frac{\beta_{10}\beta_{12}}{\beta_{10}\beta_{12} + \beta_{20}\beta_{21} + \beta_{12}\beta_{21}} \quad \text{and}$$

$$\frac{N_2}{N_0 + N_1 + N_2} = \frac{\beta_{20}\beta_{21}}{\beta_{10}\beta_{12} + \beta_{20}\beta_{21} + \beta_{12}\beta_{21}} \tag{5}$$

Thus, if aerosols are present and the recombination coefficients are finite, there will always be charged aerosols present. In the limit of pure air with $N_O = N_1 = N_2 = 0$, the small air ion density reaches an equilibrium value of $n_1 = n_2 = n_0$ where

$$n_0 = \sqrt{q/\alpha} \tag{6}$$

For a production rate of 10 ions/cm³/sec and a recombination coefficient of $\alpha = 1.6 \times 10^{-6}$ cm³/sec, the computed equilibrium ion density is $n_0 \cong 2500$ ions/cm³.[1] Under the assumed conditions, this will represent an upper limit. The mean residence time τ for ions in equilibrium is then given by:[12]

$$\tau = n_{0/q} \tag{7}$$

For the values given above the mean residence time of a small ion under these ideal conditions is approximately 250 sec.

Under normal atmospheric conditions aerosols will always be present to act as attachment sites for small ions. Under the assumption of equal recombination coefficients for both ion polarities $\beta_{10} = \beta_{20} = \beta_0$ and $\beta_{12} = \beta_{21} = \beta$ the equilibrium small ion density, n_a, with aerosols present is

$$n_a = n_0\left[\sqrt{1 + \frac{B^2}{4q}} - \sqrt{\frac{B^2}{4q}}\right] < n_0, \tag{8}$$

where $\beta = \beta_0 N_0 + \beta N_1$. For a typical rural aerosol concentration of 10^4 aerosols/cm^3 and an ion aerosol-attachment coefficient of $1. \times 10^{-6}$ cm^3/sec corresponding to 0.06 μm aerosols, the computed equilibrium ion concentration is approximately 1000 ions/cm^3 of each polarity.[13] This level is consistent with observed natural ion levels in clean air.

Small ions are removed from the atmosphere through attachment to larger particles by two different processes: diffusional charging and field charging.[13] In bipolar ion atmospheres without strong electric fields, the motion of ions is predominantly thermal or Brownian motion. This thermal motion results in collisions of the ions with aerosols in the atmosphere and the ion charge attaches to the aerosol. Diffusional charging of this nature results in an equilibrium-charge distribution described by Boltzmann statistics.

If an electric field is present as with unipolar ion atmospheres, the ions experience conductive motion along electric field lines as well as thermal motion. The field lines terminate on the aerosols. Collisions with aerosols and transfer of charge to aerosols occur because of the motion of ions along the field lines. Field charging predominates in high field regions such as are found in electrostatic precipitators.[14,15] In the presence of an electric field both charging mechanisms are present.

For aerosols with radii greater than 10^{-7}m (0.1 μm), the mean free path of the ions is less than the dimensions of the aerosol. In this case the charging process can be considered as continuous and is described by the macroscopic diffusion and conduction equation.[13]

For particles with radii smaller than 10^{-7}m the interaction of an aerosol with ions must be analyzed as individual charging events. This regime requires that the probability of a given size aerosol being charged with a given number of charges be determined. These probabilities are represented by the ion-aerosol attachment coefficients β_{ij} of Equations 3 and 4. For particles with radii less than 2×10^{-8}m no more than one elementary charge can be attached. Similarly for aerosols with radii less than 4×10^{-8}m, the charge cannot exceed two elementary charges because the strong electric field at the surface of the aerosol repels any additional charge.

Hoppel suggests a superposition of diffusional and field charging of aerosols as the best representation for combining the two mechanisms.[13] The average number of charges on aerosols is a function of the product, nt in sec/cm^3, where n is the ion concentration and t is the time. For particles with radii in the range of 1 to 4×10^{-8}m an equilibrium charge condition of 1 charge per particle is reached at nt $\cong 10^7$sec/cm^3. Hoppel notes that continuum theory gives the same result as discrete charging for nt in the range of 10^5 to 10^7.[13] Therefore, he uses the continuum for this regime which is germane to the high voltage direct current transmission line environment. An estimated time constant for attachment of small ions from a HVDC to aerosols with a concentration of 10^4 particles/cm^3 is 50 sec.[16]

III. ION CHARACTERIZATION

A. Size

The diameter and mass of air ions will, of course, depend on the number and type of molecules in the cluster. Typically, small air ions have diameters less than 2×10^{-9}m (0.002 μm). The diameter of an ion comprised of a single molecule is about 3×10^{-10}m. Fog droplets and dust particles in the atmosphere can be as large as 10^{-4}m in diameter. Most of the aerosol particles in polluted air are in the range of 10^{-8} to 10^{-7}m diameter.[15]

Perkins and Eisele have measured positive and negative ion masses at various ground level locations using a mass spectrometer.[17] Discrete values of mass were detected from 16 atomic mass units (AMU) up to 670 AMU. For one location, positive ions were seen at 18, 36, 54, 60, 72, 80, 90, 94, and 98 AMU and other masses. Positive ions were also observed in the range of 200 to 600 AMU. The masses 18, 36, 54, 72, and 90 are believed to represent protonated ammonia and its hydrates $NH_4^+ \cdot (H_2O)_n$. Negative ions were observed at 62, 80, and 125 AMU. These are tentatively identified as NO_3^-, $NO_3 \cdot H_2O$ and $NO_3^- \cdot (HNO_3)$ ions. Clearly the actual mass of a small air ion will depend on its chemical composition, which depends on the trace constituents of air and the lifetime of the ion. The measurement of ion mass at atmospheric pressure is a new technique which potentially can provide considerable information about the chemical composition of air ions.[11]

Because of the small mass of ions, they remain suspended in air. Heavier particles settle out due to gravity; for example, a 10^{-5}m diameter particle settles at a terminal velocity of approximately 0.5 cm/sec.

B. Mobility

Ions are most often classified electrically by mobility in an electric field. Mobility, k, is the drift speed, v, of an ion in a liquid or gas per unit electric field strength, E

$$k = v/E \tag{9}$$

The preferred unit of mobility is $m^2/V-s$; another commonly used unit is $cm^2/V-s$. Small air ions have mobilities in the range of 0.2×10^{-4} to $2.5 \times 10^{-4} m^2/V-s$.[3] Often the most frequently observed mobility (i.e., the mode of the mobility distribution) is reported as the average mobility. Average mobilities reported for small air ions in dry air at room temperature and atmospheric pressure are[3]

$$k_+ = 1.37 \times 10^{-4}\ m^2/V - s$$

$$k_- = 1.89 \times 10^{-4}\ m^2/V - s$$

Average values in the atmosphere are less than those measured in the laboratory because of the longer time between generation and measurement. For the atmosphere Mohnen reports STP corrected mobilities of:[3]

$$k_+ = 1.14 \times 10^{-4}\ m^2/V-s$$

$$k_- = 1.24 \times 10^{-4}\ m^2/V-s$$

Except in very high fields the mobility is independent of the electric field. This is the case in the applications being considered in this volume. Changes in gas density result in changes in collision frequency between ions and gas molecules. Consequently, ion mobility is proportional to temperature and inversely proportional to pressure. The reduced mobility k_r, represents the mobility corrected to standard temperature and pressure conditions. It is given by

$$k_r = k\left(\frac{P}{P_o}\right)\left(\frac{273}{T}\right) \tag{10}$$

where P is the atmospheric pressure, P_o is the pressure at sea level and 273K, and T is the temperature in °K.

The mobility spectra produced by a radioactive ion source and a corona ion source for several different conditions of temperature, humidity, and ion drift time were measured and compared for both negative and positive ions.[18] The mobility spectra of ions produced by

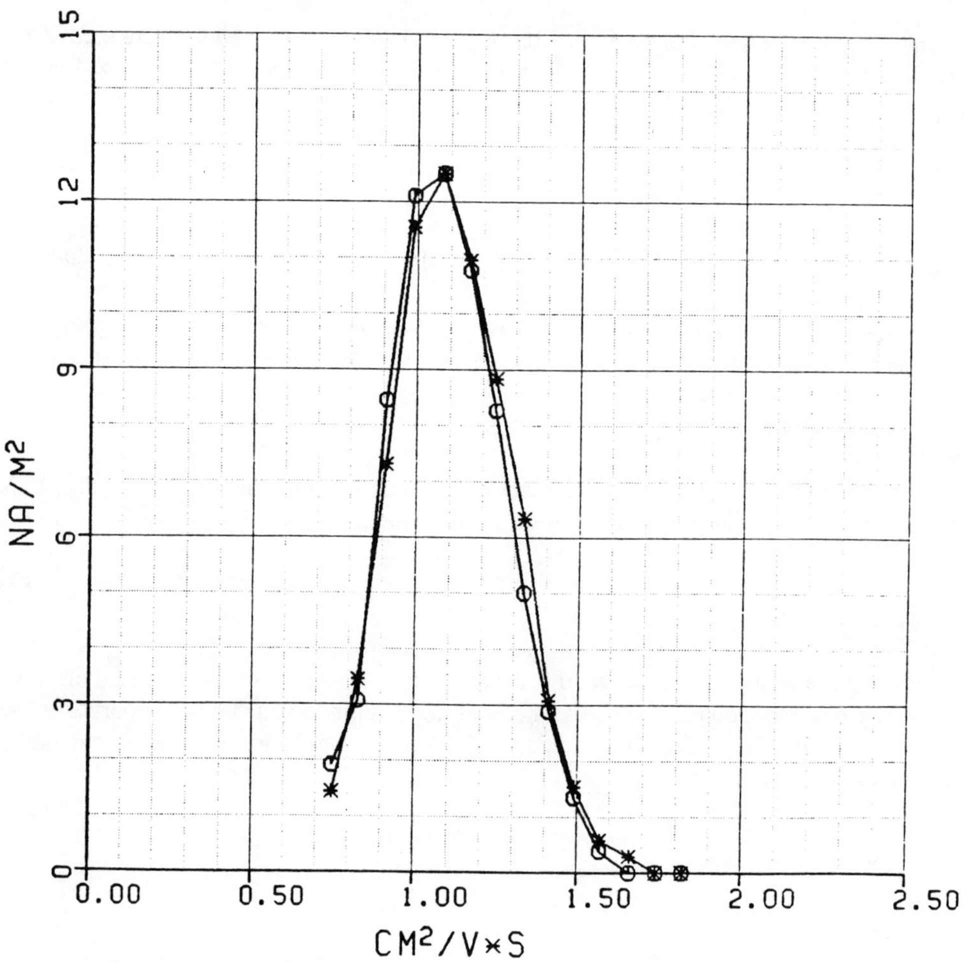

A

FIGURE 1. Comparison of mobility spectra of ions produced by corona (O) and radioactive (*) sources. (A) Positive ions: T = 20C, humidity = 55%, drift time = 0.34 s; (B) Negative ions: T = 22C, Humidity = 54%, drift time = 0.41 sec. (From Johnson, G. B., *Design, Construction, and Testing of a D.C. Bioeffects Enclosure for Small Animals,* Rep. EA 4189 Electric Power Research Institute, Palo Alto, Calif., 1985.)

these sources were identical when measured under the same environmental conditions as shown in Figure 1. Mohnen also lists average mobilities of small ions produced in "pure" air by various investigators. The results encompassing reports over a 68-year time span indicate that the average mobility is very similar for various sources: α, β, and γ radiation and corona.[3]

Mobility spectra for positive and negative ions are different as shown in Figure 2.[19] The mobility distribution of negative ions in clean laboratory air covers a broader range of mobilities than that of positive ions.[3] Spectra reported by Johnson show a similar trend, although the negative ion spectrum exhibits a narrow peak at the higher mobilities.[18] In this case, the air was not filtered and measurements of mobility were made less than a few seconds after ion generation. Because of the short lifetime, trace constituents in the atmosphere would not have interacted fully with the ions. The sharp peak in the negative ion spectrum begins to decrease and the negative ion mobility spectrum begins to fill out as the ions age. This is demonstrated in Figure 2 where the drift time is about eight times as long as that in Figure 1B.

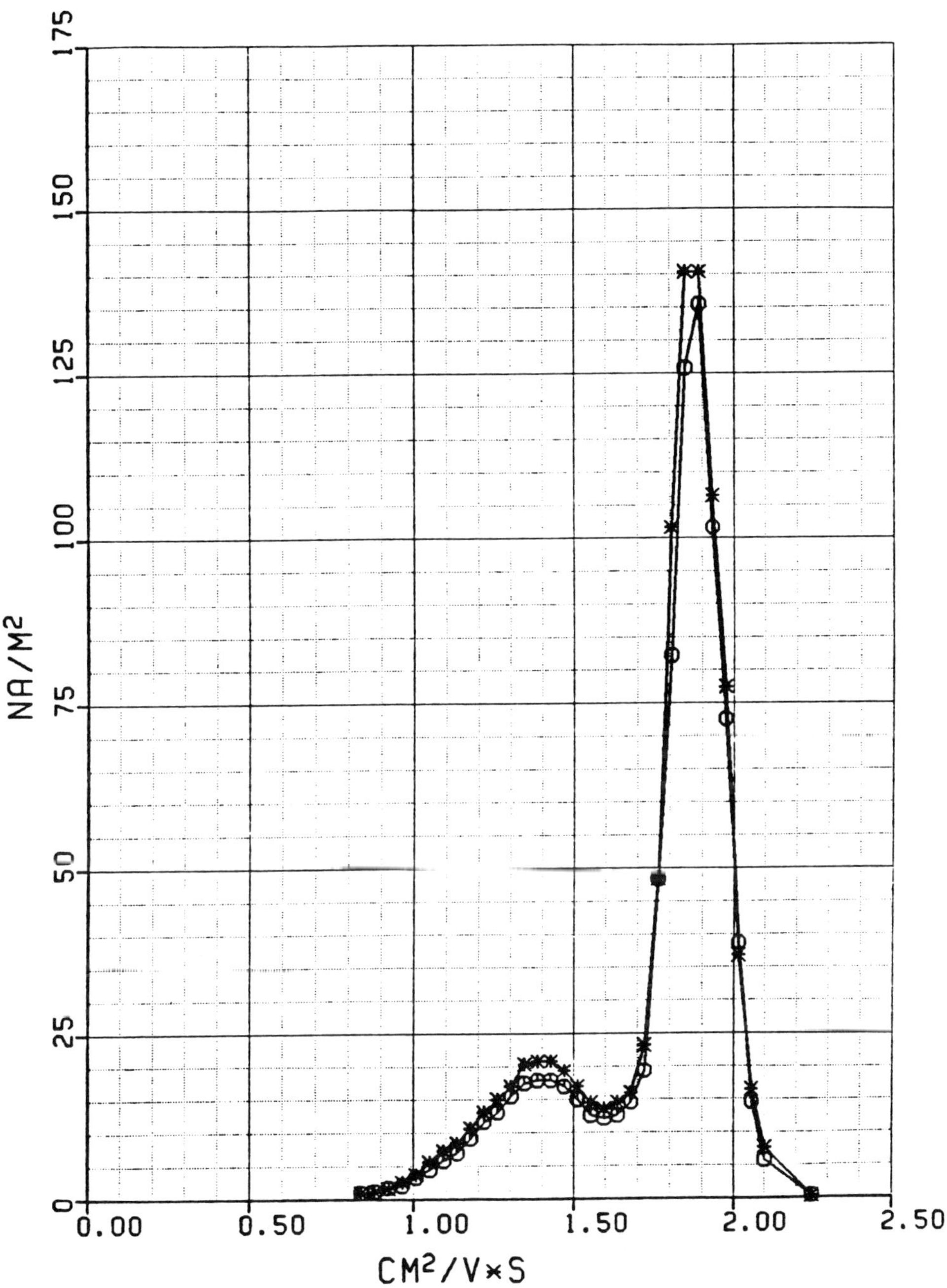

FIGURE 1B

The time required for small positive and negative air ions to reach a chemical and hydration equilibrium in a simple atmosphere at standard pressure has been calculated.[9,10] This evolution time for positive ions is less than 10^{-3} sec for absolute humidities of 7.5 g^{-3} (RH = 49%, T = 18 C) and less than 10^{-2} sec for a tenfold reduction in humidity from this. Huertas et al.[10] calculate ionic evolution times for negative ions in simple atmospheres of less than 10^{-5} sec for absolute humidities of 7.5 g^{-3} (RH = 49%, T = 18 C). Huertas and Fontan[9] point out that for times longer than 10^{-3} sec ionic evolution is due to the clustering of minor compounds other than water.

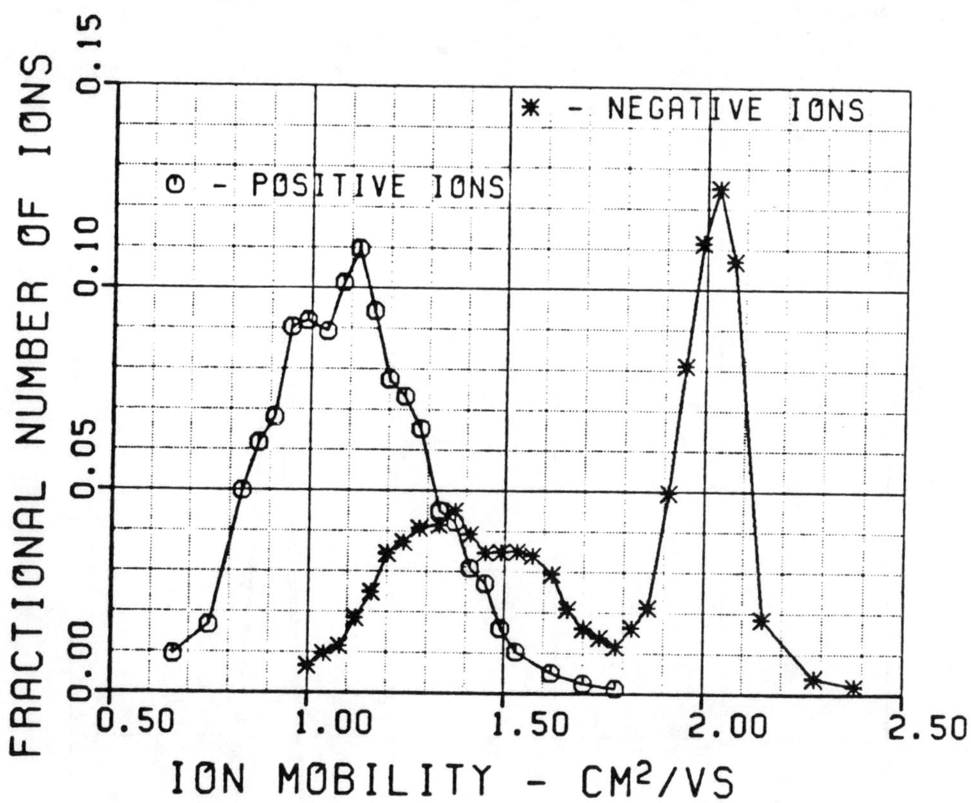

FIGURE 2. Mobility spectra for positive and negative ions. Temperature, 20°C; relative humidity, 22%; and absolute humidity, 4.5 g/m^3. The positive ions had aged about 3 sec and the negative ions had aged about 2 sec. (From Johnson, G. B. and Zaffanella, L. E., *Proc. 4th Int. Symp. High-Voltage Engineering,* National Technical University, 1983.)

Mobilities of corona-produced ions have been investigated as a function of lifetime (0.03 to 4.0 sec) and humidity (5.1 to 23.7 g/m^3) by Fujioka et al.[20] The ranges of lifetime and humidity are appropriate to the laboratory and transmission line environments discussed in later chapters. Positive ion mobility seemed to decrease only slightly with lifetime in the range of 0.03 to 4.0 sec.

Mobilities of negative ions were observed to be independent of lifetime in the range of 0.02 to 3.0 sec. Mobilities for both polarities decreased as absolute humidity increased. These observations are consistent with hydration equilibrium essentially being reached in times shorter than the observation times in these experiments.

IV. ION COMPOSITION

A. Chemical Reactions

The chemical evolution of air ions depends on the composition of the atmosphere and especially on the trace gases that are present. Air ions are continually changing as they collide with molecules in the atmosphere. The type of ion-molecule reaction which can occur depends on ionization potential and electron affinity, the proton affinity, the dipole moment and the polarizability as well as the reactivity of the molecule.

In air the primary positive ions N_2^+, O_2^+, N^+, and O^+ are rapidly converted to protonated hydrates $H^+ \cdot (H_2O)n$. The free electrons created during ion formation are quickly (microseconds) attached to oxygen, O_2^-. These initial forms then rapidly evolve through collisions and chemical reactions.

Reaction schemes which follow the evolution of positive and negative ions in simple atmospheres have been described.[3,9,10] For example, the oxonium ion and its hydrates, $H_3O^+ \cdot (H_2O)n$, are the only positive ions in equilibrium in a simple gas system containing N_2, O_2, CO_2, and traces of H_2O. In such a simple atmosphere with no molecular disassociation, two negative ions are in equilibrium: $O_2^- \cdot (H_2O)n$ and $CO_4^- (H_2O)n$. However, the possibility of molecular disassociation and the presence of trace constituents in the real atmosphere complicates reactions considerably. The above-mentioned equilibrium ions should only be considered as possible precursors of the actual ions in the atmosphere. Trace gases such as NH_3, NO, NO_2, N_2O, HNO_3, CO, CH_4, and unsaturated hydrocarbons are present as well as H_2S, SO_2, H_2SO_4, O_3, I_2, C, HCL, OH, O, and others. A collision with one of these molecules can change an ion's composition. Since the lifetimes of ions are quite long, collisions with all impurities are probable and the composition of ions is undergoing constant change. For a typical reaction rate constant of 10^{-16} m^3/s, Mohnen reports reaction times of ions with trace molecules of 4×10^{-4} and 4×10^{-1} sec for trace concentrations of 1 ppm and 1 ppb, respectively.[3] Thus, ion chemistry will continually change through chemical reactions, molecular rearrangements, and growth of molecular clusters.

Recent mass spectrometer measurements of air ions at ground level have identified the following equilibrium ions: $NH_4^+(H_2O)_n$ with n ranging from 1 to 5, NO_3^-, $NO_3^- \cdot (H_2O)$ and $NO_3^-(HNO_3)$.[17] There were also positive ions at atomic masses which could not be uniquely identified. Keesee and Castleman have speculated that the most likely candidate for a positive ion with atomic mass 60 is $(CH_3)_3NH^+$, trimethylamine, while coal tars are possible candidates for ions observed at atomic masses of 80 and 94.[21] The advent of a mass spectrometer capable of operation at atmospheric pressure offers the possibility of identification of the chemical composition of small ions.

B. Hydration

The degree of hydration will depend on the standard free energy change for hydration of the ion by one or more water molecules. Keesee and Castleman provide the standard free energy change for hydration of possible ions.[21] The peak in the hydrate distribution for ammonium is expected at n = 5: $NH_4^+ \cdot (H_2O)_5$. For the organic molecule, $(CH_3)_3NH^+$ (trimethylamine), which is the most likely candidate for the 60 AMU ion, hydration with three water molecules is expected. For the negative ion $NO_3^- \cdot (H_2O)n$, a cluster of three water molecules is expected based on free energy considerations. These degrees of hydration are consistent with the mass spectrometer observations.[17]

REFERENCES

1a. **Coulomb, C.,** *Hist. Acad. R. Sci. Paris, Mem. Math. and Phys.*, 612, 1785. Cited in Hoppel.[2]

1b. **Chalmers, J. A.,** *Atmospheric Electricity*, Pergamon Press, Oxford, 1967.

2. **Hoppel, W. A.,** The Ions of the Troposphere: Their Interactions with Aerosols and the Geoelectric Field, Ph.D. dissertation, The Catholic University of America, Washington, D.C., 1968.

3. **Mohnen, V. A.,** Formation, nature, and mobility of ions of atmospheric importance, in *Proc. 5th Int. Conf. Atmos. Electr.*, Dolezalek, H. and Reiter, R., Eds., Verlag, Darmstadt, Germany, 1, 1977.

4. **Anderson, I.,** *Mucociliary Function in Trachea Exposed to Ionized and Non-Ionized Air*, Academisk Boghandel, Arhus, Denmark, 1971.

5. **Wilkening, M.,** Characteristics of atmospheric ions in contrasting environments, Reprinted from preprint, 7th Int. Conf. Atmos. Electr., American Meteorological Society, Boston, 1984.

6. **Loeb, L. B.,** *Fundamental Processes of Electrical Discharges in Gases*, John Wiley & Sons, New York, 1939.

7. **Loeb, L. B.,** *Electrical Coronas — Their Basic Physical Mechanism,* University of California Press, Berkeley, 1965.
8. *Transmission Line Research Book: 345 kV and Above, 2nd ed.,* Electric Power Research Institute, Palo Alto, Calif., 1982.
9. **Huertas, M. L. and Fontan, J.,** Evolution times of tropospheric positive ions, *Atmos. Environ.,* 9, 1018, 1975.
10. **Huertas, M. L., Fontan, J., and Gonzales, J.,** Evolution times of tropospheric negative ions, *Atmos. Environ.,* 12, 2351, 1978.
11. **Eisele, F. L.,** Direct tropospheric ion sampling and mass identification, *Int. J. Mass Spectrom. Ion Process.,* 54, 119, 1983.
12. **Israel, H. and Israel, G. W.,** *Trace Elements in the Atmosphere,* Ann Arbor Science Publishers, Ann Arbor, Mich., undated.
13. **Hoppel, W. A.,** *Study of Drifting, Charged Aerosols from HVDC Lines,* Rep. EL 1327, Electric Power Research Institute, Palo Alto, Calif., 1980.
14. **Melcher, J. R., Sachar, K. E., and Warren, E. P.,** Overview of electrostatic devices for control of submicrometer particles, *Proc. IEEE,* 65, 1659, 1977.
15. **Friedlander, S. K.,** *Smoke, Dust, and Haze: Fundamentals of Aerosol Behavior,* John Wiley & Sons, New York, 1977.
16. **Bailey, W. H., Bissell, M., Brambl, R. M., Dorn, C. R., Hoppel, W. A., Sheppard, A. R., and Stebbings, J. H.,** *A Health and Safety Evaluation of the ±400 kV DC Powerline,* Minnesota Environmental Quality Board, St. Paul, 1982.
17. **Perkins, M. D. and Eisele, F. L.,** First spectrometric measurements of atmospheric ions at ground level, *J. Geophys. Res.,* 89, 9649, 1984.
18. **Johnson, G. B.,** Air Ion Generation and Characterization, Appendix C, *Design, Construction, and Testing of a D.C. Bioeffects Enclosure for Small Animals,* Rep. EA 4189, Electric Power Research Institute, Palo Alto, Calif., 1985.
19. **Johnson, G. B. and Zaffanella, L. E.,** Techniques for measurements of the electrical environment created by HVDC transmission lines, *Proc. 4th Int. Symp. High-Voltage Engineering,* National Technical University, 1983.
20. **Fujioka, N., Tsunoda, Y., Sugimura, A., and Arai, K.,** Influence of humidity on variation of ion mobility with lifetime in atmospheric air, *IEEE Trans. Power Appar. Syst.,* PAS-102, 911, 1983.
21. **Keesee, R. G. and Castleman, A. W., Jr.,** Ions and cluster ions: experimental studies and atmospheric observations, *J. Geophys. Res.,* 90, 5885, 1985.

Chapter 2

SMALL AIR ION ENVIRONMENTS

T. Dan Bracken and Gary B. Johnson

TABLE OF CONTENTS

I. AMBIENT ENVIRONMENTS

Naturally occurring small ion levels near ground level vary over a wide range. Fair weather values over land are typically in the range of 500 to 1000 but can easily vary from 200 to 3000 ions/cm^3 of both polarities. Because of this variability, it is difficult to give a mean value. Results of several measurements of ambient ion concentrations during fair weather conditions are given in Table 1. The ratio of positive to negative ion density near ground level is usually in the range of 1.1 to 1.3.

Parameters that affect ambient small ion levels are aerosol content of the air, weather, local geology, earth surface conditions, and altitude. The measurements in Table 1, in general, reflect fair-weather clean-air conditions remote from any local ion source. In urban areas with air pollution, small ions attach to aerosols resulting in lower small ion densities except near local sources. Equilibrium levels are determined by small ion production rates, air pollution (aerosol) levels, and aerosol-ion attachment coefficients. Typical levels in urban environments remote from local sources would be at the lower end of the values shown in Table 1 or even less.

Wind direction can affect natural levels: downwind from urban areas and other sources of air pollution aerosol densities will be higher, resulting in a reduction of small ion densities. Air coming off the ocean, where ion production rates are low, contains lower ion densities than air moving over large land masses.

During rainfall and thunderstorms small ion densities increase. Norinder and Siksna[12] report increased small ion levels at ground level during thunderstorms. Increases of negative ion densities up to 14,000 ions/cm^3 and of positive ions to above 7000 ions/cm^3 were observed. Robinson and Dirnfeld[3] observed a doubling of small ion density during rainfall. Increases during the rain were from 900 to 1500 positive ions/cm^3 and from 800 to 1700 negative ions/cm^3. The rain washed aerosols out of the atmosphere and generated negative ions by the breaking up of water (the Lenard effect). Thus, the ratio, $n+/n-$, went from 1.24 in normal conditions to 0.9 during rain. Melting snow also introduces negative charge into the air.[13]

Diurnal and seasonal variations in ion densities depend strongly on local conditions when opposite trends have been seen in different locales.[14] However, Israel reports some general observations.[15] Air ion densities reach a maximum in the early morning hours with lower values during the day as aerosol levels rise. Seasonally, summer variations are higher than winter. Over oceans, both diurnal and seasonal variations are minimal.

Local geological conditions can influence ion production rates. The most striking example of this is from measurements made in Carlsbad Cavern, N.M.[10] In this case, not only are natural radioactivity levels high but the air is confined. Wilkening measured 670,000 ions/cm^3 of each polarity in the cavern during the summer.[10] The ion production rate in the cavern was 2300 ion-pairs/cm^3 compared with a normal ground level rate of 10 to 15 ion pairs/cm^3. The increased ion production rate is due to a high concentration of radon in the cave: 65 pCi/1 vs. 0.13 pCi/1 for an outdoor site in the Rio Grande Valley of New Mexico.

Ambient small ion concentrations will be higher over granite and other minerals that contain radioactive materials than over sedementary rock such as sandstone. Near the surface of the earth the electrode effect influences the relative levels of small ions. The electrode effect refers to the tendency of a charged object to repel ions of the same polarity. Near the object's surface this results in an excess of ions of polarity opposite to that of the object. The earth has a negative charge and thus positive ion densities are enhanced near its surface by the electrode effect.

In the absence of local sources, indoor levels of small ions are determined by the presence of radioactivity from the earth or construction materials and by loss mechanisms. Removal of indoor small air ions occurs through recombination, attachment to aerosols, objects and

Table 1
MEASURED AMBIENT ION DENSITIES AT REPRESENTATIVE LOCATIONS DURING FAIR WEATHER

Location	n (+), ions/cm³	n (−), ions/cm³	Ref.
Boston, Mass.	210—400	180—345	1
Upsala, Sweden	700—1925	600—2350	2
Haifa, Israel	700—1500	575—1100	3
Bozeman, Mont.	770	520	4
France	220	180	5
England	50—2000	50—2000	6
Wisconsin	1030	790	7
Minnesota	380—800	40—1000	8
Georgia	300—400		9
Colorado	2000		
New Mexico	540	440	10
Minnesota	500	360	11

walls, and through ventilation. Indoor levels are within ranges observed naturally in the lower troposphere.[16] Highest levels in a room occur when the room is unoccupied and there is no ventilation. Aerosol production associated with human movement and activity reduces small air ion levels. Smoking causes marked decreases in small air ion concentrations. The lighting of only one cigarette in a room can cause reductions to levels of 10 to 100 ions/cm^3.[16]

Under circumstances where building materials contain relatively large concentrations of radioactive materials such as uranium and thorium, ionization rates are higher and indoor levels can exceed levels commonly experienced outdoors. For example, a room with plastered granite stone walls exhibited positive concentrations of up to 10,000 ions/cm^3. In this case, the granite contained inclusions of torbenite (copper uranite).[16]

Schreiber and Peyrous[5] describe a mountain observatory where ion levels varied daily between 1000 and 38,000 positive ions/cm^3 and between 1600 and 30,000 negative ions/cm^3. They speculate that the high levels were caused by radioactive gases which were trapped under the snow and then infiltrated into the building.

Ions released by indoor combustion sources such as gas appliances or candles can increase indoor ion densities to levels exceeding those commonly found outdoors.

II. LOCALIZED SOURCES

Small ions can be generated by various local sources, either intentionally or unintentionally. The result is a localized increase in the density of ions.

Anderson cites several examples of localized ion sources including: flame ionization near burning gas jets with ion concentrations of approximately 1×10^5 ions/cm^3; apparatus producing UV radiation such as sun lamps and electric welding with ion concentrations of approximately 1×10^4 ions/cm^3; X-ray apparatus; and radioisotopes used for scientific, medical, industrial and consumer purposes.[17] This latter group includes smoke detection, antistatic devices and room air ionizers. Water droplet breakup, such as occurs in waterfalls and showers, is another common source capable of increasing local ion densities.

Measurements near flames are reported by Johnson.[18] Candles in a basement room increased ion densities from less than 1000 ions/cm^3 up to 27,600 ions/cm^3. About 0.3 m above a burning match, ion densities of 200,000 to 300,000 ions/cm^3 were reported. However, as Hoppel pointed out, small aerosols associated with combustion processes can be charged by small ions and these charged combustion nuclei could be counted as ions.[19]

Corona discharge ionizers are used to control static electricity in many industries. Recent applications utilize a bipolar space charge from such sources to minimize spark discharges in semiconductor manufacturing processes. For one such system, ion densities of about 35,000 ions/cm^3 of each polarity were measured in the workspace of a laminar-flow clean room facility.[20] Levels in excess of 100,000 ions/cm^3 of each polarity were observed at about 1.7 m above the floor. By neutralizing charge on dust particles, this ionization system also significantly reduced the level of dust deposition on surfaces.

Room ionizers are also marketed as air cleaning devices for office and residential use. Both unipolar and bipolar devices are available. The actual ion density near such devices will depend on distance, location of the device and the particular design. For example, Johnson reported a level of 26,000 ions/cm^3 at a distance of 1.2 m from a negative ion generator.[18]

III. HVDC TRANSMISSION LINES

High voltage direct current (HVDC) transmission lines represent a line source of small ions. In recent years considerable effort has been expended in characterizing the ions near such transmission lines.[11,21-25] The electrical environment of a HVDC transmission line is complex involving varying electric and magnetic fields, ion densities and ion currents.

A bipolar DC transmission line operates with a constant voltage with respect to ground on each of two sets of conductors, or poles. The voltages on these conductor bundles are of opposite polarity with respect to ground: one positive and one negative. Constant voltages are maintained on the line, and a direct current flows through the conductors.

The electric field at the surface of transmission line conductors (the conductor surface voltage gradient) is responsible for corona occurring at that location. Corona is the partial electrical breakdown of air into charged particles.

For a particular conductor, the higher the voltage, the higher the conductor surface gradient and the more intense is the corona. The smaller the radius of curvature of an object, the higher will be the electric field at the surface of the object for a given voltage. Thus, smaller conductors go into corona at a lower voltage than large conductors; and protrusions such as water drops, dust, and insects on a conductor go into corona before the conductor itself. Operating HVDC transmission systems usually have corona present: the protrusions on conductors serve as a continuous source of small ions.

As discussed previously, primary ions, including those generated by corona, evolve in less than a second from singly charged individual molecules to the singly charged molecular clusters called small ions. Thus, in the transit time of a few seconds from the HVDC conductors to the ground, small ions have evolved to a fully hydrated condition.

Because the conductors have a constant polarity, ions with the same polarity as the conductor move away from that conductor, i.e., negative charges move away from the negative pole and positive charges move away from the positive pole. Consequently, the interelectrode space of a HVDC transmission line is filled with space charge. Most of the ions generated at the conductors migrate to the opposite pole where they are neutralized by recombination or by contacting the conductors. However, there is a significant fraction (about 10%) of the small ion current which migrates to ground or away from the transmission line.[27]

Generally small air ions are the most prevalent form of corona-generated space charge at locations on the right-of-way of a HVDC transmission line. At greater distances, the charge tends to be carried on aerosols.[18,28]

The presence of corona-generated small ions makes the electric field and charge environment complex and dynamic. The ions near the conductors are subject to a force caused by the electric field; they are also subject to convective transport by the wind. As the wind increases, more charge is transported away from the line, and the magnitudes of the electrical parameters near the line change: on the upwind side of the line, little or no space charge is present, whereas on the downwind side of the line, space charge of the same polarity as the downwind conductor is present. The line configuration also influences field and charge related effects: reducing the conductor height above ground increases field and ion effects at ground level under the conductors; and the size and number of conductors chosen and their surface condition determine the amount of corona activity. The surface condition of the conductors is strongly dependent on both weather conditions and season of the year with water droplets, insects, dust, and vegetative matter acting as the principal sources of corona. Weather conditions that affect the level of corona activity, and, thus, ground level ion densities, are precipitation, relative humidity, and temperature.

Since weather can affect corona generation and wind can strongly influence the distribution of ions, the magnitudes of the electrical parameters vary greatly with meteorological conditions. The complexity of the electrical environment makes interpretation of measurements difficult, and leads to statistical descriptions for corona phemonena, including ion effects. Due to seasonal and meteorological fluctuations, exact predictions of the distribution of corona-generated space charge near a HVDC transmission line are not possible. Instead, approximate estimates of electric field and ion levels near HVDC lines are made from analytical calculations combined with past measurements on similar lines.

There are two modes of transport for space charge from HVDC transmission lines: by electrical forces, and by convection. A small air ion with a mobility of 1.4×10^{-4} $m^2/v \cdot s$ travels at a speed of approximately 1.4 m/s in a 10 kV/m field. For a charged aerosol with mobility 10^{-6} m^2/Vs the velocity in the same field would bc 0.01 m/s. The relative importance of the two methods of transport for small ions depends on proximity to the transmission line and on wind speed. Underneath a line, conductive motion in the electric field dominates for small ions, provided the wind is not too strong. Away from a line, the earth's natural field of about 130 V/m has only a slight effect on small ions: a force comparable to a wind of only 0.02 m/s. Thus, conductive migration of small ions is very slow under the influence of the natural terrestrial field. Furthermore, at great distances from the transmission line, the electrical charge is much more likely to reside on aerosols, and the low mobility of these aerosols essentially eliminates the effect of the electric field on charge transport. In this case, turbulence and wind patterns determine dispersion of space charge.

Measurements of ion densities near HVDC transmission lines have been conducted primarily at locations directly under the conductors and at distances within usual transmission line rights-of-way. Direct measurements of ion densities are usually made at a few locations in conjunction with more numerous electric field and ion current measurements. Ion densities can also be determined from measurements of electric field and ion currents using the constitutive relationship

$$n = j/ekE \tag{1}$$

where n is the ion density, j is the measured current density, e is the electron charge, E is the measured electric field, and k is an assumed ion mobility.

Lateral profiles of measured values for electric field and ion current density near a ±400 kV HVDC test line configuration identical to that of the CU line in Minnesota are shown in Figures 1 and 2.[25] These values agree well with levels observed under the actual line in Minnesota.[25] Higher field and ion current levels than those in Figures 1 and 2 were observed when the line height was reduced from 15.2 m to 10.7 m. This latter value corresponds to the vertical clearance of the CU Line in North Dakota.

Small ion densities found directly under existing HVDC transmission lines range from ambient levels with no corona present up to levels above 100,000 ions/cm³ under certain

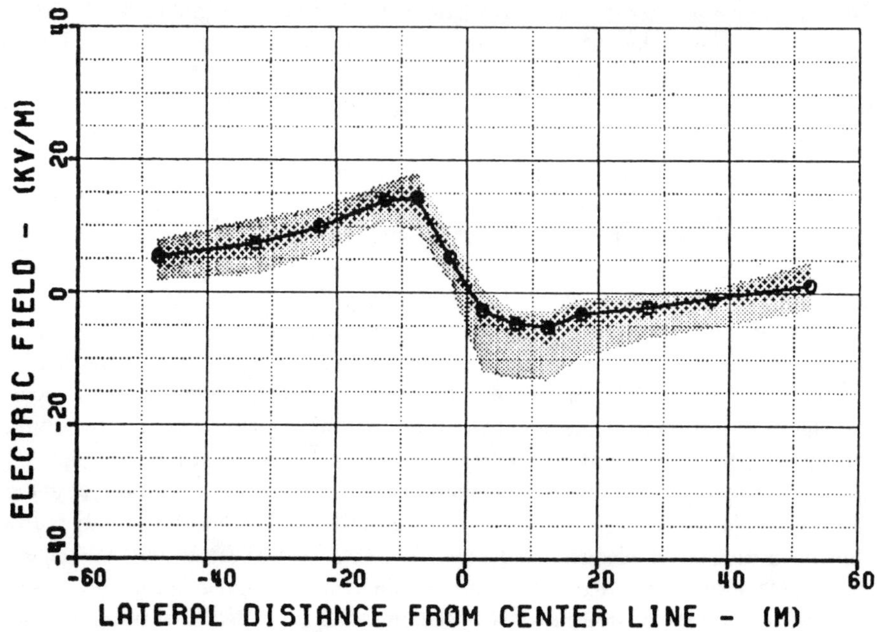

FIGURE 1. Fair weather electric field profile during October 1981 for ±400 kV test line with CU line configuration: pole spacing 12.2 m, height 15.2 m. Median level ○○; 25/75% level ▩; 5/95% level ▦.[25]

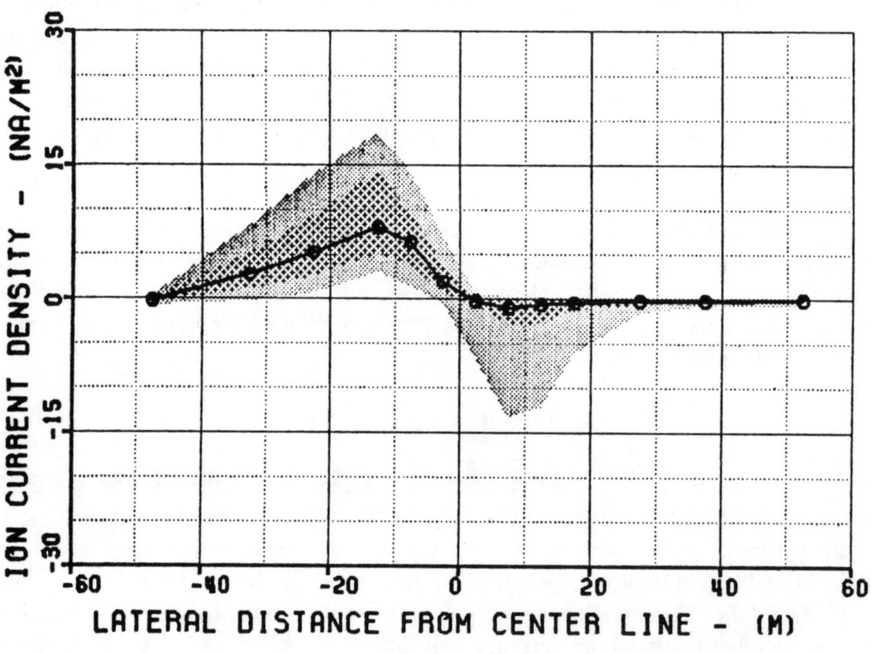

FIGURE 2. Fair weather ion current density profile during October 1981 for ±400 kV test line with CU line configuration: pole spacing 12.2 m; height 15.3 m. Median level ⊖⊖; 25/75% level ▩; 5/95% level ▦.[25]

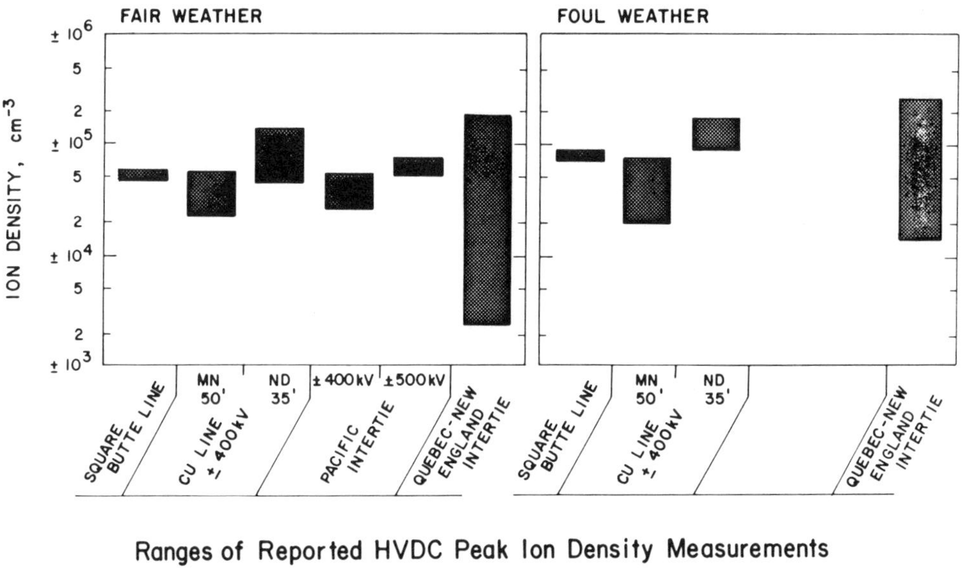

FIGURE 3. Ranges of reported HVDC transmission line peak ion density measurements.[29]

conditions. The ranges of reported peak ion density measurements under several operating or proposed HVDC configurations are shown in Figure 3.[29]

Peak foul weather levels tend to be higher than fair weather levels. However, foul weather levels also tend to fall off faster with distance from the line than fair weather levels.

Ion density measurements have been fewer and less reliable than for the other electrical parameters. Therefore the accuracy of the ranges shown in Figure 3 are more limited than similar estimates for electric field and ion currents. Nevertheless these results support useful conclusions regarding peak small ion densities near HVDC transmission lines. The maximum ion densities measured under the HVDC transmission lines considered here are less than 3×10^5 ions/cm^3 under both fair and foul weather conditions for both polarities. Typical levels under fair weather conditions with corona present are less than 10^5 ions/cm^3. Under the low corona conditions experienced during fair weather on lines with large clean conductors, ion levels are on the order of a few times 10^3 ions/cm^3 and approach ambient levels.

Measurements made at the High Voltage Transmission Research Center in Lenox, Mass., have shown that the ion density levels fall off quickly with distance.[18] Typical levels at 50 m downwind are less than 10,000 ions/cm^3. Measurements at 100 m downwind from a ± 450 kV test line (Vermont configuration) yielded small ion densities of 5000 ions/cm^3 or less even with a wind of 5 to 10 m/s. Under these conditions, the ion density under the nearest conductor to the measurement location was typically greater than 80,000 ions/cm^3. Measurements of the net charge density at the same location indicated that additional charge resided on charged aerosols.

A 3-year measurements program investigated field and ion levels at distances up to 1 mile from the ± 400 kV CU line in Minnesota.[11] Evidence of space charge from the line was measured out to 1575 m from the line where electric fields above ambient and depleted ion densities of the minority polarity (upwind conductor) were observed. However, median ion density levels at 600 to 700 m downwind from the line were less than 800 ions/cm^3 for both polarities. Maximum ion densities measured at this distance were less than 1600 ions/cm^3 for both polarities. Thus, although space charge effects from the line could be detected at these distances, the ion densities were still within the range of normal ambient variations. At 300 m downwind from the CU line, mean reported ion levels were 2100 and 3100 ions/cm^3 for positive and negative ions, respectively. The long-term mean ion density at particular locations would depend on the percent of time that the site was downwind or upwind of the line.

ACKNOWLEDGMENTS

Chapter 2 was co-authored by Drs. Bracken and Johnson. Dr. Johnson wishes to include the following acknowledgment:

The technical and financial support provided by the Department of Energy and the Electric Power Research Institute for DC electric field and ion research at the High Voltage Transmission Research Center is gratefully acknowledged. The encouragement, instruction, advice, and support of the staff of HVTRC and of my friends in the DC field and ion community are also gratefully acknowledged. Without all of the above, my participation and contribution in this book would not have been possible.

REFERENCES

1. **Yaglou, C. P., Benjamin, L. C., and Choate, S. P.,** Changes in ionic content of air in occupied rooms ventilated by natural and by mechanical methods, *Heat., Piping Air Cond.*, 3, 865, 1931.
2. **Norinder, H. and Siksna, R.,** The measurements of the variation characteristics of the density of small-ions, *Arkiv Geofysik,* Band 1, nr 1, 1, 1949.
3. **Robinson, N. and Dirnfeld, F. S.,** The ionization state of the atmosphere as a function of the meterorological elements and of various sources of ions, *Int. J. Biom.*, 6(2), 101, 1963.
4. **Sharp, E. L.,** Relation of air ions to air pollution and some biological effects, *Environ. Pollut. 3,* 227, 1972.
5. **Schreiber, G. O. S. and Peyrous, R.,** Some air ion measurements made in the Pyrenees (France) at altitudes of 200 m and 2,970 m, *Int. J. Biom.*, 23(2), 123, 1979.
6. **Hawkins, L. H.,** The influence of air ions, temperature and humidity on subjective wellbeing and comfort, *J. Environ. Psychol.*, 1, 279, 1981.
7. **Hawkinson, T. E. and Barber, D. E.,** The industrial hygiene significance of small air ions, *Am. Ind. Hyg. Assoc.*, 42, 759, 1981.
8. **O'Brien, M. V.,** Air Ion Concentrations, presented at the *76th Ann. Meet. Air Pollution Control Assoc.*, Atlanta, June 19 to 24, 1983.
9. **Perkins, M. D. and Eisele, F. L.,** First spectrometric measurements of atmospheric ions at ground level, *J. Geophys. Res.*, 89, 9649, 1984.
10. **Wilkening, M.,** Characteristics of atmospheric ions in contrasting environments, in *Proceedings of VII Int. Conf. Atmospheric Electricity,* American Meteorological Society, Boston, 1984, 54.
11. **Hendrickson, R. C.,** Electrical Environment Outside the Right of Way of CU-TR-1, Report 5, Power Plant Siting Program, Minnesota Environmental Quality Board, St. Paul, 1985.
12. **Norinder, H. and Siksna, R.,** Ionic density of the atmospheric air near the ground during thunder-storm conditions, *Ankir Geofysik,* Band 1, nr 16, 453, 1951.
13. **Muir, M. S.,** Electric space charge in the atmosphere, *J. Atmos. Terrestrial Phys.*, 34, 941, 1972.
14. **Knoll, M., Eichmeier, J., and Schon, T. E.,** Properties, measurement, and bioclimatic action of small multimolecular atmospheric ions, in *Advances in Electronics and Electron Physics,* Marton, L., Ed., Academic Press, New York, 1964.
15. **Israel, H.,** Atmospheric Electricity, Vol. 1, Fundamentals Conductivity, ions, Israel Program for Scientific Translations, National Science Foundation, Washington, D.C., 1971.
16. **Reiter, R.,** On the presence and generation of AC and DC electric fields and small ions in closed rooms as a function of building materials, utilization, and electrical insulation, *J. Geophys. Res.*, 90, 5936, 1985.
17. **Anderson, I.,** *Mucociliary Function in Trachea Exposed to Ionized and Non-ionized Air,* Academisk Boghandel, Arhus, Denmark, 1971.
18. **Johnson, G. B.,** The electrical environment and HVDC transmission lines, in *Conf. on Environ. Ions and Related Biological Effects,* Charry, J. M., Ed., American Institute of Medical Climatology, Philadelphia, 1982, 66.
19. **Hoppel, W. A.,** 1982. Discussion of [18].

20. **Unger, B. A., Chemelli, R. G., and Bossard, P. R.,** A room ionization system for electrostatic charge and dust control, in *Electrical Overstress/Electrostatic Discharge Symp. Proc.,* Philadelphia, 1984, 40.
21. **Hill, H. L., Capon, A. S., Ratz, O., Renner, P. E., and Schmidt, W. D.,** *Transmission line reference book: HVDC to ±600 kV,* Electric Power Research Institute, Palo Alto, Calif., 1977, 73.
22. **Bracken, T. D., Capon, A. S., and Montgomery, D. V.,** Ground level electric fields and ion currents on the Celilo-Sylmar ± 400 kV DC Intertie during fair weather, *IEEE Trans. Power Appar. Syst.,* PAS 97, 370, 1978.
23. **Comber, M. G., Nigbor, R. J., and Zaffanella, L. E.,** HVDC Transmission Line Research, Report EL-2419, Electric Power Research Institute, Palo Alto, Calif., 1982.
24. **Maruvada, P. S., Dallaire, C. D., Norris-Elye, O. C., Thio, C. V., and Goodman, J. S.,** Environmental effects of the Nelson River HVDC transmission lines RI, AN, electric field, induced voltage and ion current distribution tests, *IEEE Trans. Power Appar. Syst.,* PAS 101, 951, 1982.
25. **Johnson, G. B.,** Electric fields and ion currents of a ± 400 kV HVDC test line, *IEEE Trans. Power Appar. Syst.,* PAS 102, 2559, 1983.
26. **Bracken, T. D.,** Electrical parameters of the high-voltage direct-current transmission-line environment, in *Biological Effects of Extremely Low Frequency Electromagnetic Fields,* Proc. 18th Annu. Hanford Life Sci. Symp., Phillips, R. D. et al., Eds., Technical Information Center, U.S. Department of Energy, Washington, D.C., Conf-781016, 485, 1978.
27. **Sarma, M. P., and Janischewskyj, W.,** Analysis of corona losses on D-C transmission lines: II-bipolar lines, *IEEE Trans. Power Appar. Syst.,* PAS 88, 1476, 1969.
28. **Bailey, W. H., Bissell, M., Brambl, R. M., Dorn, C. R., Hoppel, W. A., Sheppard, A. R., and Stebbings, J. H.,** *A Health and Safety Evaluation of the ±kv DC Powerline,* Minnesota Environmental Quality Board, St. Paul, 1982.
29. **Banks, R. S., Charry, J., Bracken, T. D., Berkin, E., Marshall, J. H., and Flynn, M. D.,** The Public Health Implications of HVDC Transmission Facilities: An Assessment of the Available Evidence, New England Electric Transmission Corporation and Vermont Electric Power Company, 1982.

Chapter 3

MEASUREMENT OF ELECTRIC FIELD AND ION-RELATED QUANTITIES*

Ronald H. McKnight

TABLE OF CONTENTS

I. INTRODUCTION

Measurements of various quantities to describe the electrical characteristics of the atmosphere have been made for many decades by atmospheric scientists using a variety of instruments and measurement techniques. The purpose of this chapter is to describe those techniques which have application in present-day ion-related research. References chosen are representative and it is not intended that they be exhaustive. Summaries, specialized texts,[1,2] and proceedings of conferences[3] are excellent sources for more detailed discussions of particular topics and should be consulted by the interested reader.[4]

The electrical properties of the troposphere are highly variable, but it is worthwhile to review typical values of some of the parameters associated with the unperturbed atmosphere. In many respects, the troposphere acts as a weak plasma. Ionization is caused primarily by radioactivity in the air and ground, and by cosmic rays. Ion densities are typically of the order of 3×10^8 ions/m^3 for both positive and negative ions while the magnitude of the electric field at the surface of the earth is of the order of 100 to 200 V/m. These average values depend on location, and may be significantly perturbed during inclement conditions and display diurnal and other time dependent variations.

In addition to the natural background sources, the electrical atmosphere may also be locally modified by human activities. High voltage transmission lines produce electric fields near the lines which are much larger than those found naturally. For high voltage dc (HVDC) transmission lines, the electric fields at the surface of the conductors are usually large enough to produce a corona discharge in the air surrounding the conductor. This corona discharge produces appreciable ion densities at ground level near the line and can result in measurable amounts of space charge being present downwind from the line.[5,6] Earlier experiments carried out by atmospheric scientists studied the dispersal of a plume of charge produced from an elevated wire in corona discharge and showed that charge can be transported significant distances from its localized sources.[7] More recently the dispersal of plumes of ions has been used to study air flow around structures.[8] Similarly, the use of electrostatic air cleaners in air handling systems can produce unwanted space charge in interior spaces while the use of ionizers in industrial settings to control static electricity can result in a change in the indoor ion density.[10] There are also situations where dc fields and ions may be introduced into exposure facilities to study their effects on biological systems.[11-14]

II. DEFINITION OF TERMS

To avoid confusion in the following sections, it is useful to provide some concise working statements of terms which are frequently used. This discussion assumes familiarity with elementary college-level electricity and magnetism theory.[15]

When dealing with slowly varying fields, as is the case in the present discussion, the electric field E is generally defined to be the negative gradient of the electric potential U. In much of the earlier work published in atmospheric electricity, however, the positive direction of the electric field was defined as pointing into the earth, denoting the field as the positive, rather than the negative gradient of the electric potential.[16] A similar problem in definition has emerged in recent publications describing electric fields and ion currents associated with HVDC transmission lines, where plots of field values show positive values of electric field in a direction opposite to that indicated by the geometry of the line.[6] This convention results in a presentation of results which is opposite to that based on the accepted definition of electric field and may cause confusion to the unwary reader.

Primary ionization processes in the atmosphere separate one or more electrons from a molecule or atom which then becomes an ion. In times less than 1 sec, these species undergo a large number of reactions, i.e., "age", and result in what are normally called small air ions.[17] Laboratory investigations indicate small ions are, in fact, molecular ions which have formed clusters with other molecules, with water molecules playing an important role in the ion aging process.[18,19] These small ions will continue to react, in a complicated way, with trace gases in the atmosphere and with other ions. The motion of ions in an electric field is described in terms of their mobility k, which is defined as the speed per unit electric field. In general, atmospheric ions (with physical sizes of about 0.001 μm) have a range of mobilities, which for small ions is of the order of 0.1 to 2×10^{-4} m²/Vs.[18] Small ions can attach to larger particles, in which case there is a significant decrease in the mobility. These larger particles, which are predominately in the 0.01 to 0.1 μm size range, have mobilities which are orders of magnitude smaller than those of small ions, in the range of 0.3 to 1×10^{-6} m²/Vs.[20,21] The number distribution of ions and charged particulates as a function of mobility is sometimes referred to as the mobility spectrum f(k), and the number of ions with mobilities between k and k + dk is given by dN = f(k) dk.

The presence of ions and electrons results in space charge, which is free charge existing in the atmosphere. In atmospheric physics, it is conventional to refer to space charge as the free unbalanced charge, while space charge density ρ is a measurable quantity defined as

$$\begin{aligned} \rho(\mathrm{C/m^3}) &= \sum_i n_i q_i \\ &= \sum_i n_i^+ q_i - \sum_i n_i^- q_i \end{aligned} \tag{1}$$

where n_i is the number of ions per m³, with charge q_i (positive or negative), and the sum extends over all the free charge composing the space charge in a given volume. For purposes of the present discussion, the terms space charge density and net space charge density will be used interchangeably. The unipolar (or polar) positive or negative space charge density can also be measured. Frequently, net space charge density values are converted to an equivalent ion density value by assuming that the ions making up the space charge are singly charged, an assumption which is justifiable for small ions but is less certain for charged particles. Since net space charge density only measures the imbalance between positive and negative charge in the air, it is possible to have large numbers of positive and negative ions (large polar space charge density), but a small net space charge. Because of the presence of charge carriers, air is conductive. This conductivity, λ, is related to the ions making up the space charge through the equation

$$\lambda = \lambda_+ + \lambda_- = \sum_i n_i q_i k_i = \sum n_i^+ \,|\, q_i \,|\, k_i^+ + \sum n_i^- \,|\, q_i \,|\, k_i^- \tag{2}$$

where the sum extends over all charge carriers of density n_i, charge q_i, and mobility k_i. Frequently, the polar conductivity for a given polarity is written as

$$\lambda \pm = \rho \pm k \pm \tag{3}$$

where $\rho\pm$ is the space charge density, and $k\pm$ represents an average mobility for the ions responsible for the conductivity.

It is clear that the primary contribution to the conductivity will be due to small ions because of their large mobility, unless the number of charged particles is orders of magnitude greater than that of the small ions, a situation which may occur in polluted air.[22] The current density $\vec{J}$ can be defined in terms of other quantities as

$$\vec{J} = \lambda \vec{E} \tag{4}$$

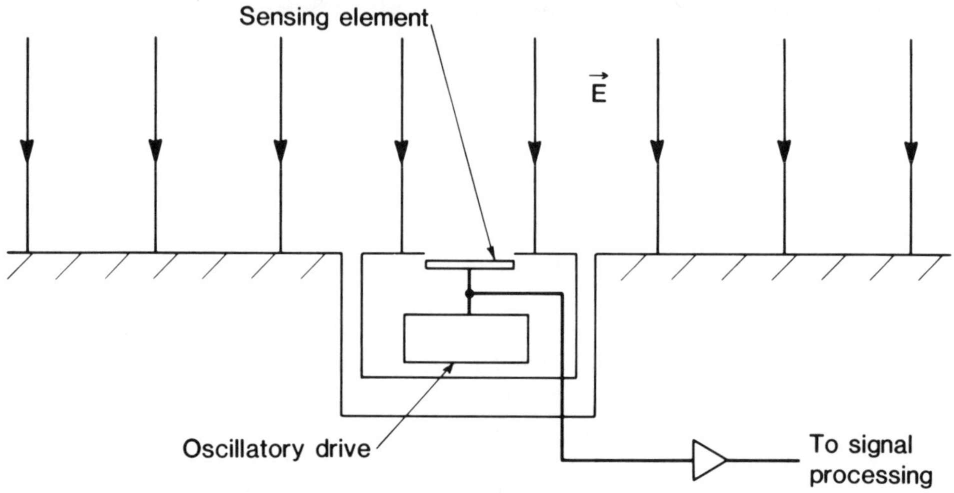

FIGURE 1. Schematic representation of vibrating sensor electric field meter mounted in ground plane. System details are omitted.

where the quantities have been previously defined. Usually, only the vertical current density is measured and this component is frequently referred to as current density.

The presence of aerodynamic forces in the form of winds or air flow in an instrument is a very important aspect of the measurement problem. Relevant aerodynamic terms are not included in the list of electrical parameters above, but they will be defined, when necessary, in the following discussions.

III. MEASUREMENT OF ELECTRIC FIELD AND SPACE POTENTIAL

A. Electric Field

A direct measurement of the electric field in or near the ground plane using instruments designed for this purpose is of most use in present-day applications. The two most commonly used instruments are the vibrating plate sensor,[23] shown schematically in Figure 1, and the field mill which is shown in Figure 2.[24] With each instrument, the electric field existing at the sensor face induces a surface charge on the active element that is proportional to the electric field. This charge is mechanically converted into an alternating signal which can be processed by standard signal handling techniques. In the case of the vibrating plate sensor, the signal results from modulating the input capacitance of the system which is coupled to the field, while for the field mill, the sensing plates are periodically covered and uncovered, resulting in a variation of the charge induced on the plates. The devices are readily calibrated for the space charge free situation by using a parallel plate geometry with the sensor element located in the ground plane as shown in Figure 3. By following guidelines regarding plate spacing and size which have been developed for ac measurements,[25] calibrations of adequate accuracy can be obtained in a straightforward way. Either instrument can be made to have sufficient frequency response to follow the tracking of field fluctuations, while both have adequate dynamic range to cover expected field values from ambient up to 100 kV/m. In addition to their use in atmospheric research, both types of instruments have been operated in the more severe electrical conditions around HVDC transmission lines, and have been used in long-term measurements.[26,27]

If space charge is present, resulting in current to the sensing element, then consideration must be given to errors which may be introduced by this current. A careful study was done

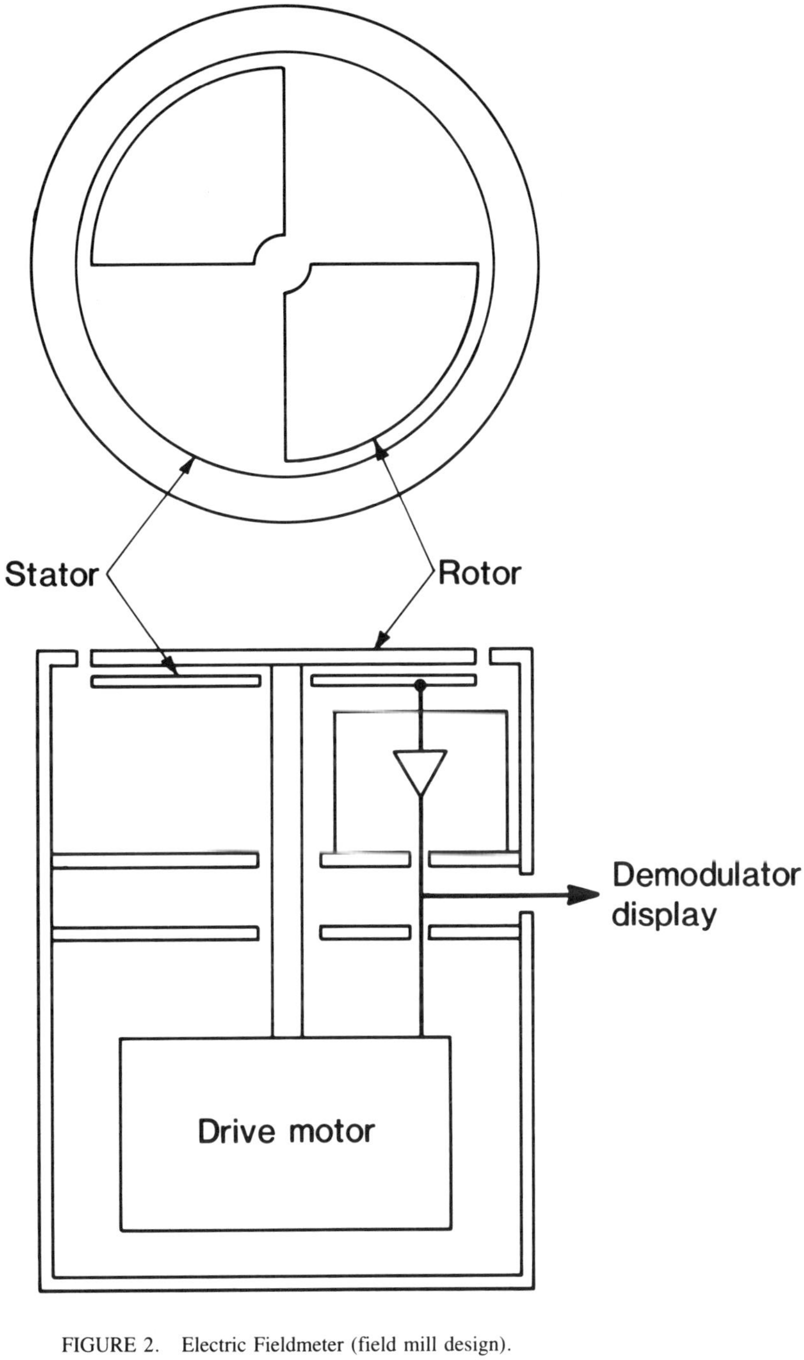

FIGURE 2. Electric Fieldmeter (field mill design).

using a specially developed parallel plate system, which produced an electric field with space charge in which the electric field was calculable.[28] This study showed that vibrating plate instruments can be subject to considerable error because of current incident on the sensor, while a field mill device can be insensitive to these currents. This characteristic of the field mill can be shown theoretically.[29,30] An example of the results obtained in this study are shown in Figure 4, which compares the calculated electric field with the measured electric field obtained from two different types of E-field meters. The observed error was small for conditions similar to those encountered near HVDC transmission lines, but the actual error depends on the size of current density and the electric field. In some exposure

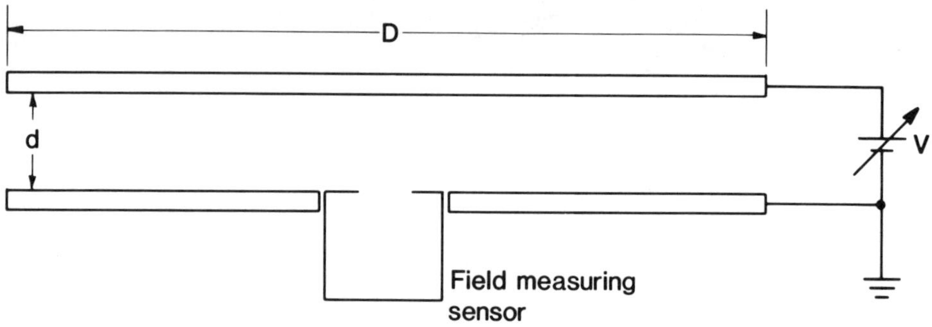

FIGURE 3. Calibration system for calibrating electric field meters.

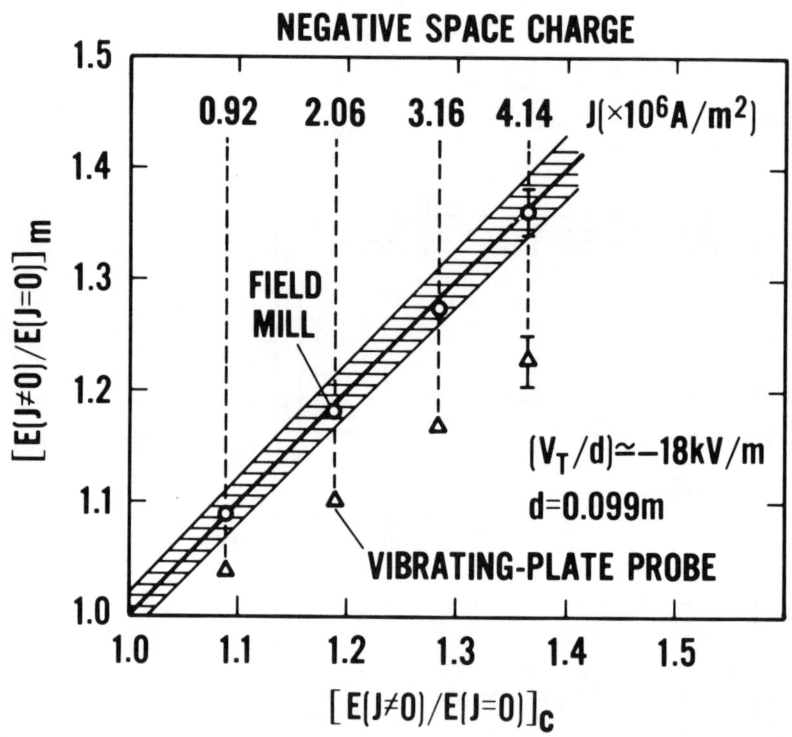

FIGURE 4. Comparison of calculated and measured electric field for vibrating plate and field mill instruments. Incident current densities (A/m^2) are indicated. Estimates of uncertainty in calculated field are shown as cross hatched area. Data are normalized to space-charge-free value. (From Misakian, M., *J. Appl. Phys.*, 52, 3135, 1981. With permission.)

systems, the results of this study may not be directly applicable, since the incident current densities and electric field values are outside the range of values considered. Descriptions of specialized devices for use in determining electric fields at the surface of conductors in corona, where space charge is present, may be found in the literature,[31,32] and a recent review article discusses applications of field mills to specialized measurements of electric fields.[33]

Instruments used outside the laboratory, such as those to monitor the long-term performance of a dc transmission line, are usually subjected to severe environmental conditions which may affect the operation of the device and the accuracy of the measurements. Under

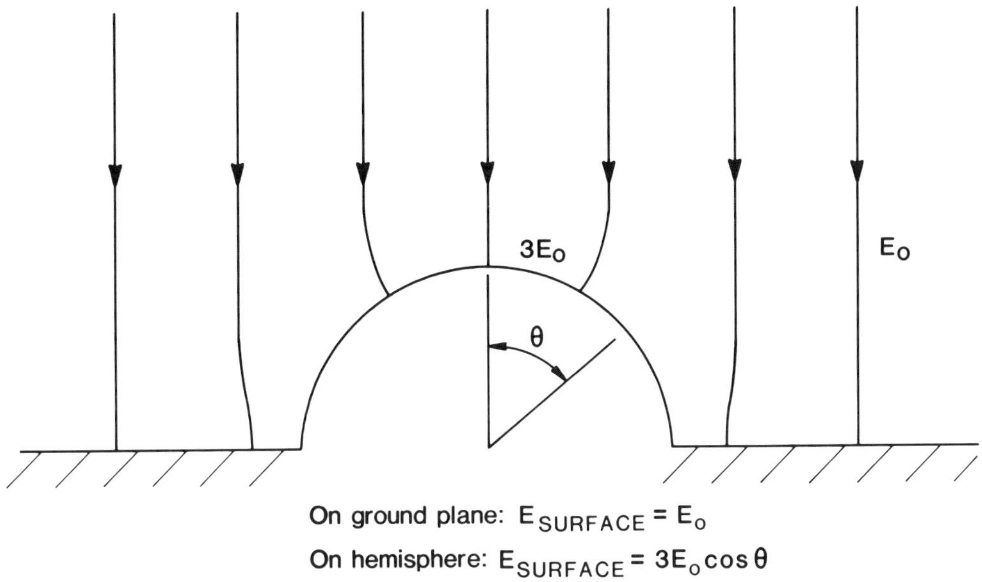

FIGURE 5. Field enhancement at surface of conducting hemispherical boss located on ground plane in a uniform field. Theoretical expression for surface charge density is indicated.

these conditions, reliable measurements will be obtained only if methods of periodically checking instrument operation, calibration, and drift are considered as part of the measurement process. Because of the possibility of charge accumulation on insulating surfaces, careful attention is necessary to ensure cleanliness. Charge accumulation can result in instability in measurements and in long-term drift. For ground-level measurements, sensors should be located flush with the ground plane, since significant errors can result from misalignment. If the sensing device is located above ground, then there is a geometric enhancement of the field which must be determined if an accurate reading is to be obtained. A simple calculable example of this phenomena is illustrated in Figure 5 for a hemispherical boss located on a ground plane in a uniform field.[15] It is unlikely that a field measuring device would have a configuration which could be represented by such a simple geometry. For actual sensors located above the ground plane, the enhancement factor would have to be determined by using a parallel plate calibration system or by comparing a ground level and elevated sensor in a known field. For such a calibration, the calibration facility would have to be of appropriate dimensions to avoid the introduction of systematic errors.[26,34] Even if the enhancement factor is determined in a parallel plate system, it may not be the same in a nonuniform field. In the case of making a measurement above the ground plane where significant amounts of space charge are present, there is the additional problem of accounting for the field due to the space charge.

An interesting device using cylindrical geometry has been designed to measure electric fields above ground in the vicinity of HVDC transmission lines.[27] This device is reported to be free of errors due to incident currents and other errors which may result from the above ground operation. The use of double field mills to determine the potential gradient above the surface of the earth has also been described.[35,36] It is important to note that in both these measurements, some method of maintaining the measuring system at a voltage equal to the space potential at the location of the device must be provided to minimize the distortion of the unperturbed field by the measuring system.

In addition to the field probes described above, there are commercially available devices, operating on various principles, which are designed to aid in the detection of fields due to static charges. These devices, which are usually hand-held and intended for use primarily in industrial environments, are not designed for accurate measurements. They are useful only for providing approximate values of existing fields. Calibration is performed with the device located a given distance from a plane electrode.

Electric field probes which use electro-optic sensors have also been developed and are described in the literature.[37] The commercial development of these devices is at an early stage and it is not clear to what extent they will become available. One clear advantage that this type of measurement system has over those described above is that the probe is electrically isolated from ground and so it could be used to provide spatial information about dc electric fields in the presence of space charge and without introducing significant perturbations. The effects of charging of the probe itself due to current flow are unknown.

B. Space Potential

Direct methods for determining the electric potential at a given spatial location (space potential) involve the use of a probe which is located in space and which is configured so that its potential relative to ground can be measured. The probe is brought to the potential of the space surrounding it by the exchange of charge mediated by ions in the vicinity of the conductor. For normal atmospheric conditions, the time required to obtain equilibrium is long, and so a local source of ions is provided, usually in the form of a radioactive source attached to the conductor. Although simple in principle, the radioactively equilibrated space potential ''antenna'' has been the subject of much discussion in the literature.[38-41]

IV. MEASUREMENT OF VERTICAL CURRENT DENSITY

If both positive and negative ions and an electric field are present above a ground plane, there will be conduction current flow to the ground. In the unperturbed atmosphere, the electric field at the surface of the earth is such that positive ions move to the earth and negative ions are driven upward. The current density associated with this current flow is a few picoamperes/m^2. Two techniques for determining the vertical current density have been used. If the conductivity and electric field are measured simultaneously at the same location, then the vertical current density can be calculated from their product. A direct measurement of the current density can be made by locating a collector plate, isolated from ground, in the ground plane as shown in Figure 6A. Historically, this type of sensor is known as a Wilson plate.[42] The average current density, J, over the sensing element is given by $J = I/A$, where I is the total measured current.

Although the measurement appears straightforward, systematic errors can affect its accuracy. If the electric field at the surface of the collecting plate changes, perhaps due to the drift of a cloud of charge overhead, then a displacement current results which is indistinguishable from conduction current. Techniques for compensating for signals which result in changes of the electric field at the surface of the sensor have been described in the literature.[43,44] These techniques involve the use of special impedance matching or simultaneous measurements of the field and a suitable subtraction of the displacement current from the total current recorded by the current plate. It is also clear that the use of appropriate time averaging, which is straightforward with contemporary data acquisition techniques, can minimize those displacement current contributions that are due to the drift of charge over the sensor. A second source of nonconduction current can result from the mechanical transport of charge to the collector plate due to turbulence of the air flow over the collector plate. The question of turbulent mixing and its role in determining the distribution of ions near the surface of the earth has been described recently.[45] Mechanical transport of charge to the collector plate is probably of little importance in measurements where the motion of ions is determined primarily by electric fields (electric forces dominate hydrodynamic forces).

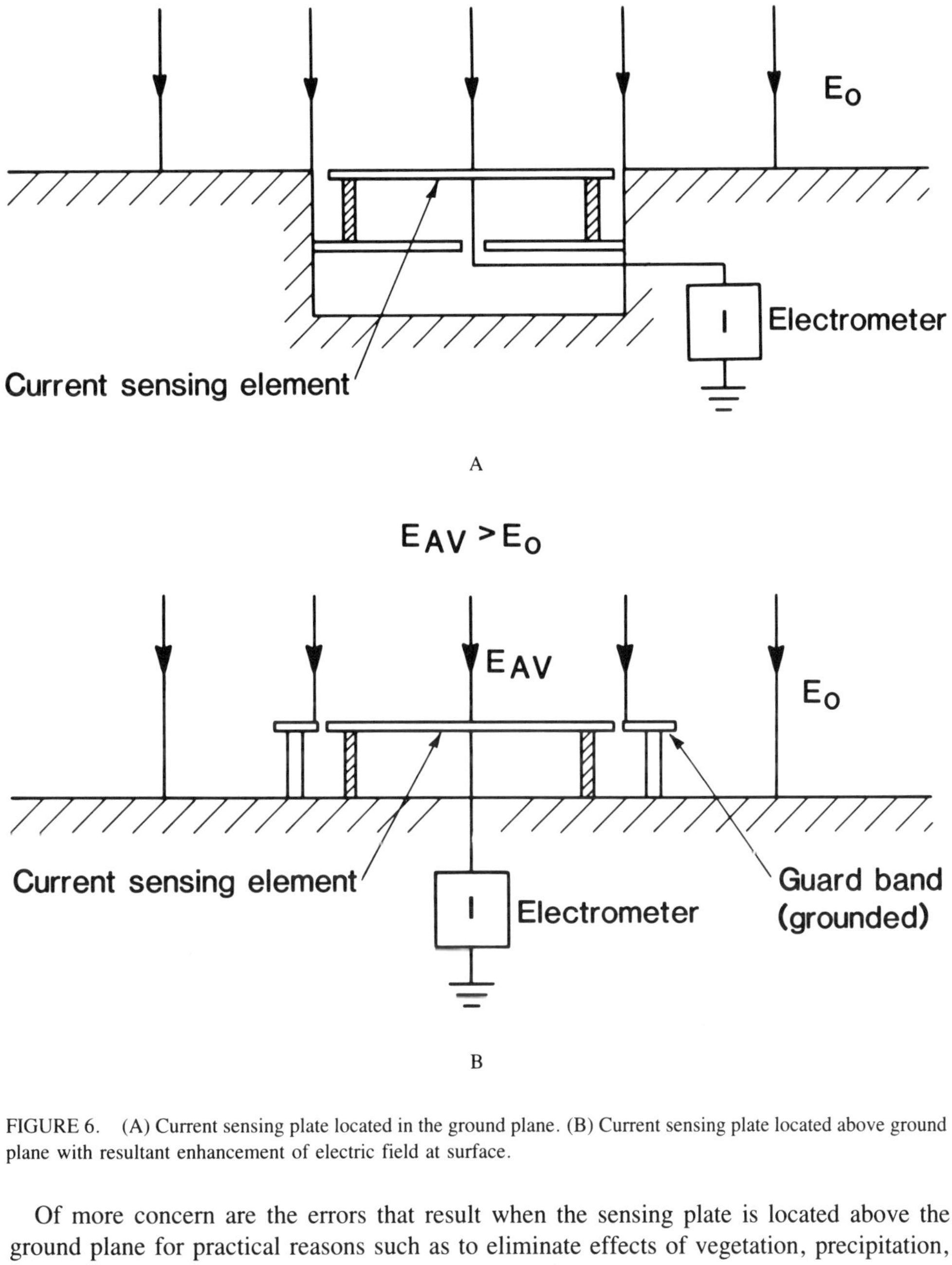

FIGURE 6. (A) Current sensing plate located in the ground plane. (B) Current sensing plate located above ground plane with resultant enhancement of electric field at surface.

Of more concern are the errors that result when the sensing plate is located above the ground plane for practical reasons such as to eliminate effects of vegetation, precipitation, or insects. In this case, there is an enhancement of the average electric field at the surface of the sensing element and a resultant increase in the indicated current, as shown schematically in Figure 6B. Even if the sensing element is surrounded by a guard band, the errors can be significant and depend on the geometry and location of the sensor.

A systematic study has been done to determine the errors associated with above ground measurements of vertical current density.[46] This investigation was done primarily in the laboratory using various model sensor geometries, but also included actual field data taken under a test line. The results of the study are summarized in Figure 7. During the course of the study, geometric enhancement values for a given model sensor configuration were determined using both ac (60 hz) techniques and dc field and ion methods. For the conditions to be expected under a HVDC transmission line, the two methods gave the same results, as

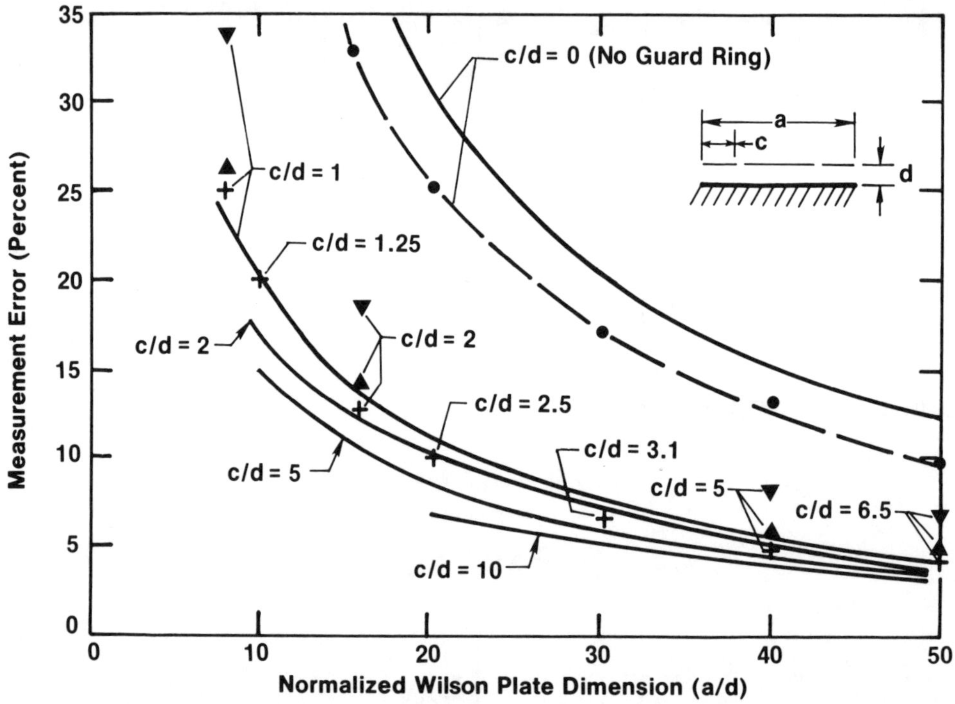

FIGURE 7. Errors for current sensors located above the ground plane. Parameters a, c, and d are indicated on the figure. Solid curves are drawn through data points obtained by using ac fields to determine field enhancements. Symbols (+, ▲, ▼, and ●) indicate data points for dc fields with ions. DC operating conditions were comparable to those found under operating HVDC transmission lines. (Additional data in McKnight, R. H., Kotter, F. R., and Misakian, M., *IEEE Trans. Power Appar. Syst.*, PAS 102, 934, 1983.)

is shown in Figure 7. In the figure, the solid curves represent composite data obtained using ac techniques, while the dc results are shown as discrete data points. In effect, then, for a given sensor configuration, the geometric enhancement factor can be determined using a parallel plate system energized with a 60-hz source[46,47] with either full size or scaled sensors.

Small current probes made from 0.16 cm-thick copper-coated circuit board are frequently used to measure current density in exposure systems.[11] These probes are useful for characterizing system current density uniformity and for monitoring long-term system performance. Even for this small spacing above the ground plane, the errors introduced are not negligible. For example, a square probe made from circuit board with a 1 cm^2 sensing element surrounded by a 1-cm-wide guard band would have a systematic error of approximately 8% if placed on top of the ground plane during the measurements. This result suggests the care which must be taken to insure that the plane of the probe and the ground plane coincide, if the probe is located in the ground plane. The actual current which impinges on a surface which may be uneven or covered with vegetation may be significantly different because of field enhancement and would have to be determined by an actual measurement. If the field at the surface is sufficiently high, there may be corona which would complicate the description of the electrical environment.

V. MEASUREMENT OF CONDUCTIVITY

As was pointed out previously, the conductivity of air can be determined by measuring, independently, the current density and electric field at a point and calculating their ratio. Conductivity may also be measured directly by using a cylindrical aspiration device known

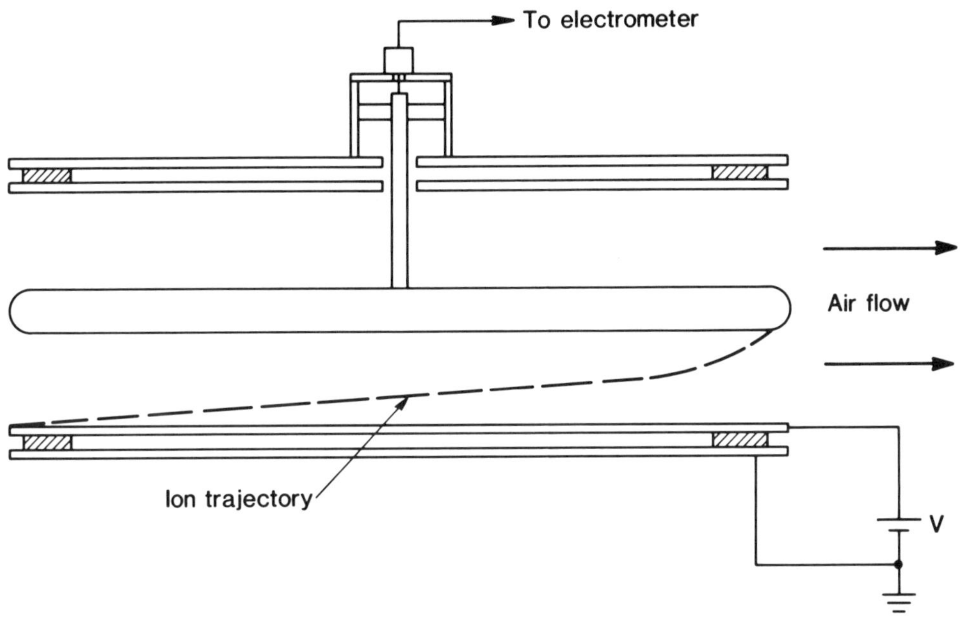

FIGURE 8. Gerdien tube (schematic). Outer grounded cylinder is used to limit extent of fringing fields (edge effect).

as a Gerdien tube[48] which is illustrated in Figure 8. In the configuration shown, ions move under the influence of the air stream, and the electric field established by the polarizing potential. Some fraction of the ions is deposited on the inner cylinder. Either positive or negative ions are collected, depending on the polarity of the polarizing potential, and the unipolar conductivity associated with these is measured. Figure 9 indicates, schematically, the current-voltage (I-V) characteristics of a Gerdien tube for a fixed flow rate through the instrument and no fringing field effects (see below). If the voltage and current are proportional (Ohms law is satisfied), a measure of conductivity is obtained. For certain operating conditions, i.e., decreased flow rate or increased polarizing potential, a saturation condition results as indicated. By writing an expression for the current in terms of physical parameters, this behavior can be described analytically in the following way:

$$I = M_o e \left\{ \int_o^{k_c} \frac{k}{k_c} f(k)\, dk + \int_{k_c}^{\infty} f(k)\, dk \right\}$$

$$k_c = \frac{M_o \epsilon_o}{V_o C} \qquad (5)$$

where C is the effective interelectrode capacitance of the Gerdien tube, V_o is the polarizing potential, and M_o the volumetric air flow rate.[49] The critical mobility, k_c, is the limiting mobility value for those ions which are completely collected by the Gerdien tube. All ions with mobilities $k>k_c$ will be collected, while only a fraction of those with mobilities $k<k_c$ will contribute to the measured current. In this expression for the current to the inner electrode, the expression f(k) is the mobility distribution function and f(k)dk is the number of ions with mobility between k and k +dk. The charge on an electron is e, and ϵ_o is the permittivity of air. It is assumed that all ions are singly charged.

If all of the ions present have mobilities $k<k_c$, then the expression for the current reduces to

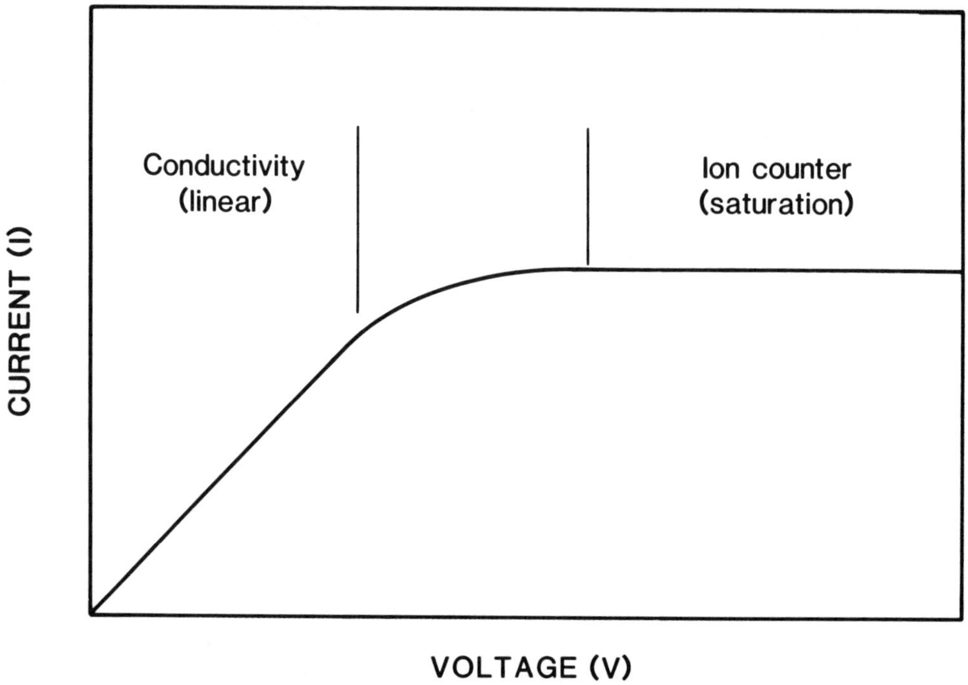

FIGURE 9. Features of I-V curve for a Gerdien tube for the operating range from conductivity measurements to ion counting.

$$I = \frac{M_o}{k_c} \int_o^{k_c} e\, k\, f(k)\, dk = \frac{V_o C}{\epsilon_o} \lambda \tag{6}$$

This expression relates the conductivity of the air drawn through the Gerdien tube and a set of physical parameters which can be determined. Note that the result is independent of the volumetric air flow through the device. For the situation where some of the ions have mobilities $k>k_c$, the response is no longer proportional to conductivity, and the I-V characteristic is nonlinear. In terms of Equation 5, the integral from k_c to ∞ begins to contribute. When all the ions are being collected, the curve saturates as shown in Figure 9. In this mode of operation, the device is said to function as an ion counter.[48] Because of the large mobility differences between small ions and charged particulates, the I-V curve of an ion counter will display a plateau when all the small ions are being collected even in the presence of charged particulates. This is technically not a ''saturated'' condition since, as the polarizing voltage is increased, a larger fraction of the less mobile particles will be collected. However, small ion densities are determined by operating the ion counter in this saturated condition. In practice, Gerdien tubes are operated with high flow rates and low polarizing potentials to assure that the device is operating on the linear part of the I-V curve. It is important that the proper operating conditions be established by determination of the I-V characteristic for the Gerdien tube being used. Although the conductivity measurement is one associated with small ions, if the air is heavily polluted the small ion density may be substantially reduced and in fact represent only a small fraction of the charged particulate population. If such conditions exist, an interpretation of the conductivity as being representative of small ions may not be warranted.

One source of systematic error associated with conductivity measurements made using the Gerdien tube results from the effects of fringing fields which are sometimes called ''edge effects''. These effects should be distinguished from those which are due to fields external

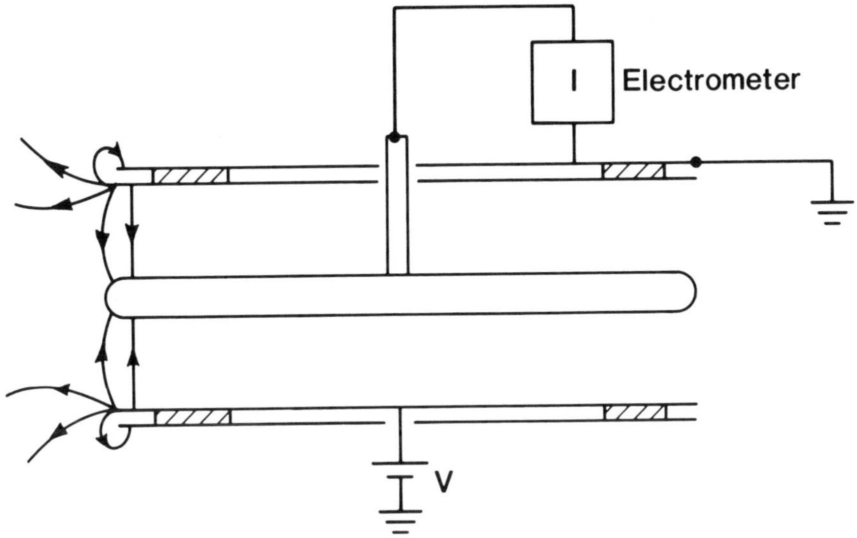

A

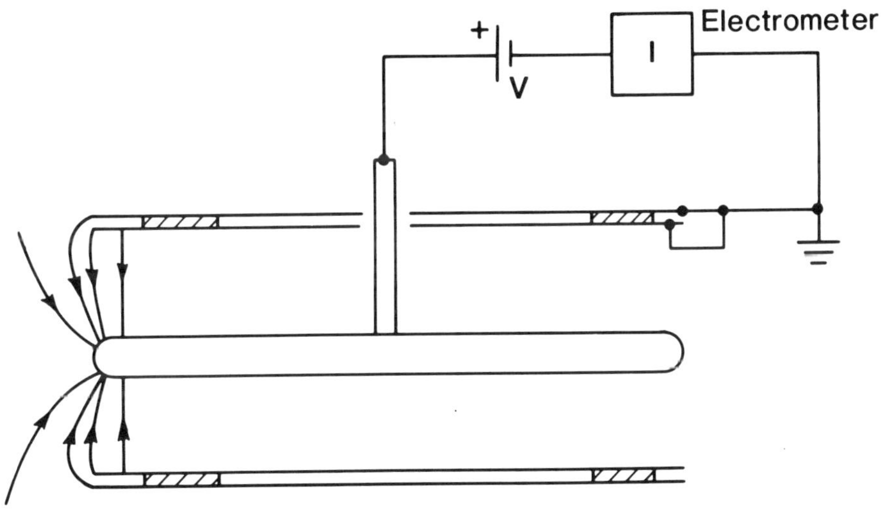

B

FIGURE 10. (A) Gerdien tube polarizing potential configuration such that ions are deflected at inlet. (B) Arrangement where fringing fields attract ions to the collector from outside the device.

to the instrument, which are discussed in a separate section. The polarizing potential may be imposed on the interior in several ways, two of which are indicated in Figure 10. If the collector electrode is grounded (through the electrometer), as indicated in Figure 10A, the polarizing potential on the outer electrode results in external fringing fields which affect the measurement by retarding and causing the loss of incoming ions. In an actual instrument design, this effect may be somewhat reduced by surrounding the outer electrode with a grounded cylinder which restricts the extent of the external field as indicated. When the polarizing potential is in series with the electrometer (Figure 10B), the external fringing field is reduced. In this case, the field can still ''leak'' to the outside and result in an increased number of ions which will be attracted to the collector rod. This condition can easily be detected by reducing the air flow through the tube to zero and noting the presence of current. For the configuration shown in Figure 10B, there are severe restrictions on polarizing supply stability and leakage across insulators, since spurious signals are directly coupled to the electrometer. These problems can be minimized in a number of ways which are discussed in the literature[50,51] and are summarized in an extensive treatise on ion counters.[52]

One practical problem is the determination of the effective capacitance appearing in Equation 5, which is that associated with the active ion-collecting region of the Gerdien tube. Methods for making this measurement are discussed in References 52 and 53.

In contrast to the situation for electric field sensors and current density probes, there exist no methods for independent calibration of ion-measuring devices such as Gerdien tubes or ion counters, primarily because there are no standard ion sources. In principle, the characteristics of these instruments are based on calculable or measurable quantities, but because of the number of potential sources of systematic errors, a means of calibration would be desirable. A recent set of experiments has been completed in which the use of the parallel plate system described above was investigated as a standard ion source, and some of the results of this investigation are summarized in a later section.[54] Small radioactive sources are useful in setting up measurements, and the individual investigator should, insofar as is possible, verify the correct operation of any measuring instrument and its ancillary equipment. There are potential sources of error which are common to all aspiration instruments, such as the effects of external fields and losses in ducts that are discussed in another section, and these must be considered when assessing the overall accuracy of a measurement of an ion-related electrical quantity.

VI. MEASUREMENT OF NET SPACE CHARGE DENSITY

A. Cage Methods

Measurement of net space charge density in the atmosphere is most readily carried out by using one of two techniques.[55] The first of these is known as the Faraday cage method and is illustrated in Figure 11. An enclosure made of a coarse conducting mesh is located in the region where the measurement is to be made. Assuming that the space charge enters the enclosure and uniformly fills it, then a solution can be obtained for Poisson's equation for the interior of the enclosure. The ease of the solution and the results obtained depend on the geometry of the cage. The solution for a spherical cage is straightforward, but such a geometry is difficult to implement. For a parallel-piped enclosure, the solution of Poisson's equation is[56]

$$\Phi(x,y,z) = \frac{64\rho}{\epsilon_o \pi^3} \sum_{\substack{i,j,k \\ \text{odd}}} \frac{\sin\left(\frac{i\pi X}{A}\right) \sin\left(\frac{j\pi Y}{B}\right) \sin\left(\frac{k\pi Z}{C}\right)}{\left[\left(\frac{i\pi}{A}\right)^2 + \left(\frac{j\pi}{B}\right)^2 + \left(\frac{k\pi}{C}\right)^2\right]} \tag{7}$$

where ρ is the space charge density in the enclosure, A, B, and C are the dimensions of the cage as shown in Figure 11, and the other constants have their usual meaning. A comparable expression can be written down for the electric field by taking the gradient of the sum in Equation 7. For a cubical enclosure, the potential expression sums to the simple form

$$\Phi = 6.32 \times 10^9 \, \rho \, L^2 \tag{8}$$

for the potential at the center of the cube.

Although the solution for Poisson's equation is more straightforward for the rectangular

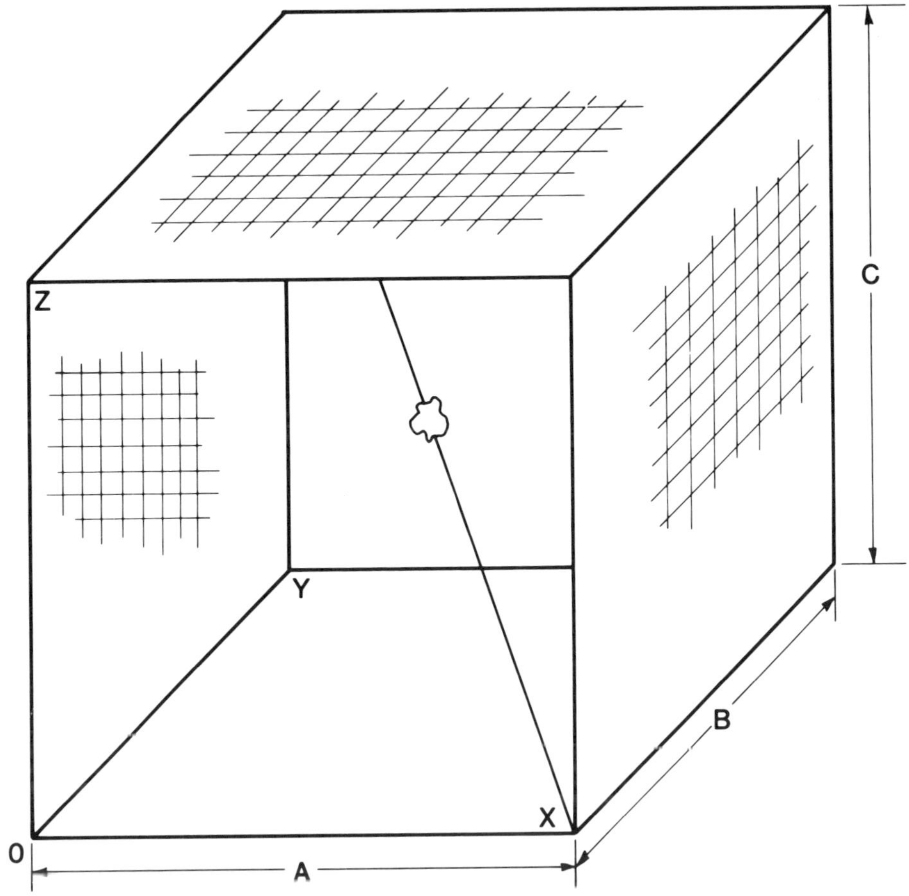

FIGURE 11. Cubical Faraday cage with equilibrated potential antenna at the center.

enclosure, a right circular cylinder presents a convenient geometry for construction, and solutions for Poisson's equation are presented in the literature.[57] If the cage is located in a high field region, the cylinder is less likely than a cube to produce corona from cage edges. The determination of the space charge density inside the cage can be made by measuring either the potential at an internal point, usually the center, or by measuring the electric field at the boundary of the enclosure. The potential at the center can be measured by using a radioactively equilibrated antenna which has been discussed previously. For a cubic enclosure 1 m on a side and a net space charge density of 1.6×10^{-9} C/m^3 (10^4 ions/m^3), the potential at the center is 10 V. For these same conditions, the electric field at the center of one of the faces is 51 V/m. Presently available electrometer devices, designed for picoampere measurements, can be used in a simple configuration to measure the potential.[58] A measurement of comparable quality of the E-field would require a meter with excellent long-term stability, low drift, and the capability of measuring small fields accurately. Such devices can be constructed, but are not readily available.

One requirement for the Faraday cage method to give an accuate measurement is that there be adequate air motion to ensure uniform distribution of the space charge throughout the enclosure and to inhibit diffusion of charge to the cage walls. The material for the enclosure wall must allow charge to move freely through the mesh, but shield the interior of the cage such that the influence of external fields is minimal.

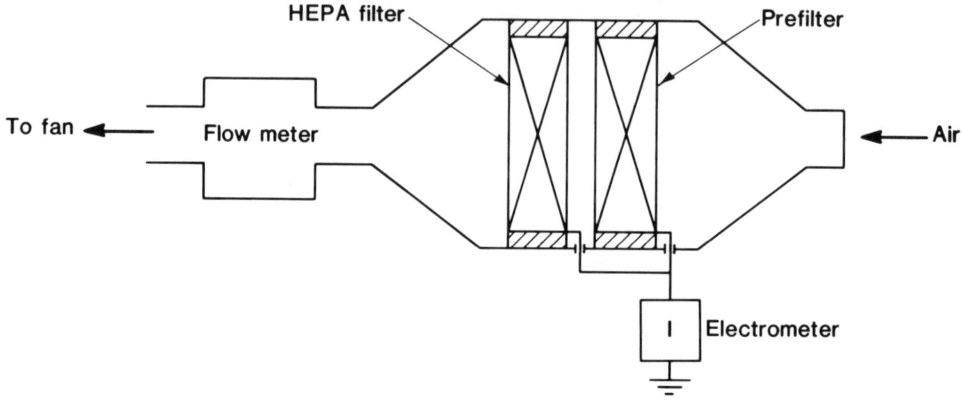

FIGURE 12. Filtration device for measuring net space charge density. Both prefilter and HEPA filter are isolated from ground.

B. Filtration Methods

A second method for measuring net space charge is known as the filtration method[59] and is illustrated in Figure 12. Here, an air sample is drawn into an enclosure with a filter through which all the air passes. The filter is electrically isolated from ground so that the current, I, resulting from the charge removed from the air stream by the filter, can be measured. From this current, the net space charge density can be calculated from $\rho = I/M_o$, where M_o is the volumetric air flow through the filter. This method of measuring space charge is accurate only if the filter collects almost all of the charge in the air stream. Different filter materials have been used in the past, including steel wool and various fibers. Using high efficiency air particulate (HEPA) filters, recent measurements have been made which indicate that only a small fraction of the incident charge passes through this type of filter.[60-62] A direct comparison between cage measurements and those made using a HEPA filter indicated that the filter method appears to be equivalent to the Faraday cage technique, to within experimental uncertainty.[56] A facility designed to produce a test volume containing space charge[63] has been used in a laboratory measurement of the transmission of a HEPA filter for a number of different space charge conditions. The results of this measurement indicate that for a space charge made up predominately of small ions, less than 0.1% of the ions pass through the filter.[64]

Each of the two methods is subject to systematic errors. The Faraday cage method is appealing since, in a sense, it is an "absolute" method, but large enclosures are required to obtain adequate signal if the space charge density is small. There are also questions about the loss of small ions to the grounded enclosure if the cage is located in high electric fields, such as may be found near a HVDC transmission line.

Filter measurements require an aspiration device and accurate determinations of both current and flow rate. HEPA filters must be used with prefilters in dirty environments to avoid clogging. The effects of external fields at the inlet must be determined, as must any losses in ducting leading to the filter enclosure. Problems relating to external field effects are common to several measuring methods and are discussed later.

One laboratory application of a filter would be to provide a comparison calibration of an ion counter. If the space charge is unipolar, then a measure of net space charge density and polar space charge density should yield the same result. Since measurements made using ion counters are usually subject to more systematic uncertainty than those made using a filter, a comparison of ion counter and filter measurements under these conditions could be used to calibrate the ion counter in terms of the filter. This comparative type of measurement has not been done.

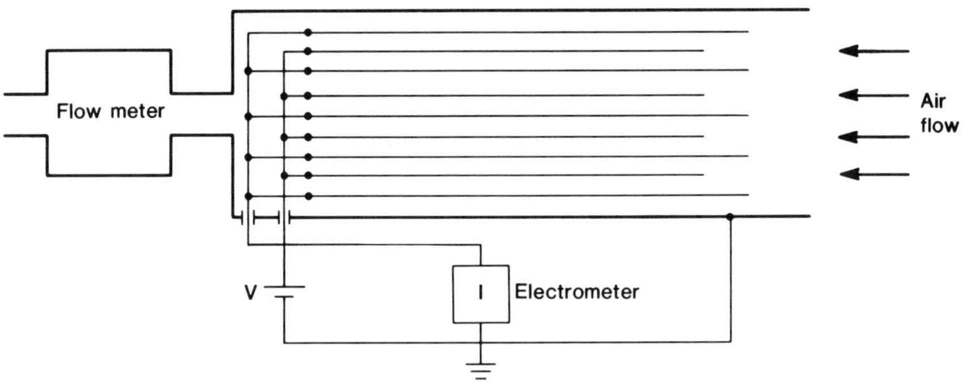

FIGURE 13. Parallel plate ion counter. Volume between outer collector plate and counter enclosure is inactive (no polarizing potential).

VII. MEASUREMENT OF UNIPOLAR ION DENSITY

Although the net space charge density may be small, the number of positive or negative charges in a region may be large. Unipolar space charge density can be measured using aspiration devices historically known as ion counters.[65] These devices, which were originally adaptations of Gerdien tubes, do not count ions but measure the polar charge density. If a Gerdien tube is operated in the saturation region, then all ions are collected, as was shown earlier. For the ions collected, the space charge density is given by $\rho = I/M_o$, where I is the current measured from the counter, and M_o is the volumetric flow rate through the counter. It is common in the literature to assume that all of the collected ions are singly charged, and to convert the measured charge density to an equivalent number of charges per unit volume, or, more simply, as statement of ions/m^3.

In addition to the cylindrical geometry, parallel plate configurations are used for ion counters as illustrated in Figure 13.[51] Alternate plates are connected and the groups of plates are connected to a polarizing potential and an electrometer, respectively. It is relatively easy to collect all the small ions because of their high mobility. An ion counter, like that shown in Figure 13, with five collector plates 46 cm long, a plate to plate spacing of 0.95 cm, a polarizing potential of 22.5 V, and a flow rate of $16.5 \times 10^{-3} m^3/s$ (35 CFM) will have critical mobility of $0.31 \times 10^{-4} m^2/Vs$. In one instrument, two sets of plates are used to allow a relative measure of large and small ions.[66] Commercial instruments are frequently configured with selectable polarizing voltages to allow operation with different critical mobilities. For charged aerosol (large ion) measurements the cylindrical geometry is used, with long cylinders, close spacings between inner and outer cylinders, and high voltages. In present-day studies of the size distributions of aerosols, adaptations of the cylindrical geometry with a merging air stream provide differential information, rather than the integral response obtained from usual operation.[67]

Ion counter measurements can be subject to many systematic errors which, for discussion, may be divided into those associated with the interior of the counter, and those external to the counter. The errors outlined earlier in the discussion of conductivity measurements are also relevant for the cylindrical geometry used in an ion counting mode and will not be repeated here. The parallel plate counter can show a significant loss of small ions if the electrode geometry is not properly designed. The polarizing plates must be sufficiently recessed behind the collecting plates, or fringing fields can precipitate ions to the enclosure walls, resulting in an effect associated with the geometry of Figure 13. An example of measurements demonstrating this effect is shown in Figure 14.[68] Here, the relative positions

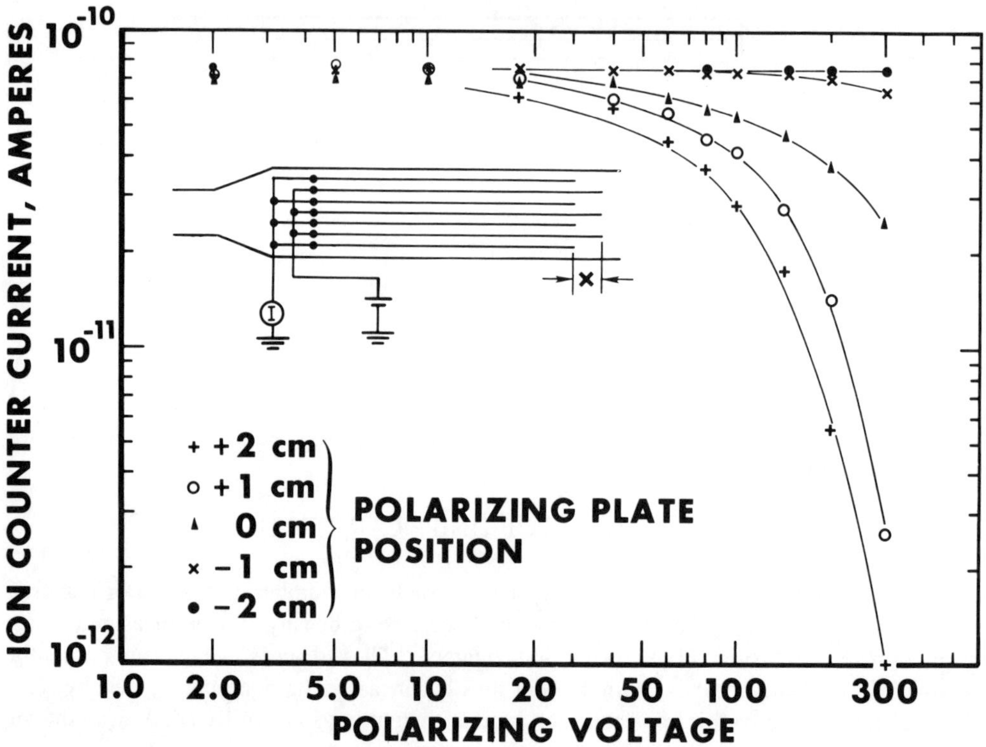

FIGURE 14. Ion counter output current as a function of polarizing voltage for different plate configurations. The displacement of the leading edges of the collection and polarizing plates is x. Positive values of x correspond to a situation where the polarizing plates protrude in front of the collector plates. For x = 0, the plates are flush. (From McKnight, R. H., Kotter, F. R., Misakian, M., and Hagler, J. N., 1981 Annual Report: Electric and Magnetic Field Measurements, National Bureau of Standards, NBSIR 82-2527, 13, 1982.)

of the collecting and polarizing plates were varied and an I-V curve was obtained for each setting. The flow rate and ion density were held constant. It is clear that as the polarizing plates move even with, or forward of, the collecting plates, there are large losses of small ions at the inlet. This effect is analogous to that shown in Figure 10A for cylindrical geometry. A useful empirical guide is that the polarizing plates be recessed two plate spacings behind the front of the collector plates, but the presence of fringing fields is best shown by experimentally determining the I-V characteristic of the counter.

Measurement of the volumetric flow rate is straightforward if a suitable flow meter is used, but this is usually not done in commercial instruments because of the cost of the flow meter. For these devices, the flow rate is determined by the manufacturer and is included in a given instrument calibration. If the counter is operated in a situation where there are external winds, the air flow through the instrument can be affected and the stated instrument calibration may become invalid. In parallel plate systems, an additional problem arises because of the inactive region existing between the outside collecting plates and the counter enclosure (Figure 13). Air flow in this region is measured as part of the volumetric air flow, but does not contribute ions to the measured current. Ideally, flow should be restricted to the active region between the plates, although if the flow profile in the counter is known, a correction can be made for this volume. The shape of the ion counter and the characteristics of the air flow through the device are also of concern, but are of more importance in determining mobility spectra than in ion counting. The usefulness of converging air flow is discussed in the literature.[49]

Vibrations of the electrodes can result in large microphonic noise signals which may obscure the actual signal if the space charge densities are small. Parallel plate counters are more susceptible to these microphonic problems than are cylindrical systems because of their construction, but for all systems, the effect depends on the individual construction and the magnitude of the vibration. Polarizing potentials should be applied by batteries, although this may not always be possible. Instabilities in the power supply will result in perturbations of the counter output, since the polarizing potential is capacitively coupled to the electrometer used to measure the current.

Sources of error due to external effects include external fields, losses in ducts carrying the air stream to the counter, and the perturbation of the volume surrounding the counter due to the air flow through the counter. The presence of space charge in the instrument can result in a change of the field distribution and affect the operation. These matters are discussed in a following section.

In industrial applications, air ionizers are used to produce increased concentrations of air ions in areas where there is concern about deleterious effects of static electricity. The effectiveness of such ionizers in reducing these effects is determined in different ways, including the use of charged plate to collect ions.[69]

VIII. MEASUREMENT OF ION MOBILITY AND SPECIES IDENTIFICATION

As indicated previously, space charge is made up of small ions and charged particles which have a distribution of mobilities. Both conductivity and polar ion density measurements use devices which have a mobility-dependent response. Unfortunately, an accurate determination of ion mobility spectra at atmospheric pressures is a difficult measurement, especially under the conditions normally encountered in the field. On approach, which is of use in obtaining a low resolution estimate of the mobility spectrum, uses the I-V characteristics of an ion counter. This aspiration method will only be outlined here, since extensive descriptions of the method may be found in the literature.[4,68] The mobility dependent response of an ion counter to space charge is given in Equation 5. Taking derivatives of the response function and rearranging, results in the following expression,

$$f(k) = \frac{C\,V_o^3}{e\,M_o^2\epsilon_o}\,\frac{d^2I}{d\,V^2} \tag{9}$$

where the terms have been defined previously. The extraction of ion mobility distributions from numerical data using this technique involves a second order numerical derivative. In practice, data may be manipulated in various ways to extract mobility information. An improvement on this method, in which the conventional Gerdien tube is constructed with a sigmented collector, is shown in Figure 15. By using this type of device, the numerical manipulation is reduced to one derivative.[71] The primary difficulty with either method is the necessity for obtaining smooth data, so that meaningful numerical derivatives can be calculated. During data acquisition, a parameter such as flow or polarizing voltage is varied. If averaging is required to reduce fluctuations in the data, then the data scan takes some time and the mobility distribution may change during the acquisition period. Examples of mobility spectra obtained using ion counter techniques[49,71,72] are shown in Figure 16 where they are compared with measurements made using a time-of-flight method that is described below.

A specialized technique using air flow, but which does away with the requirement to determine the integral response of the ion counter, is outlined in Reference 73. In this approach, ions are injected into a filtered air stream and move under the action of air flow and electric fields to a collector plate (Figure 17). This "air blast" method produces a direct measure of the mobility spectra, since it is a differential technique. This method has had a

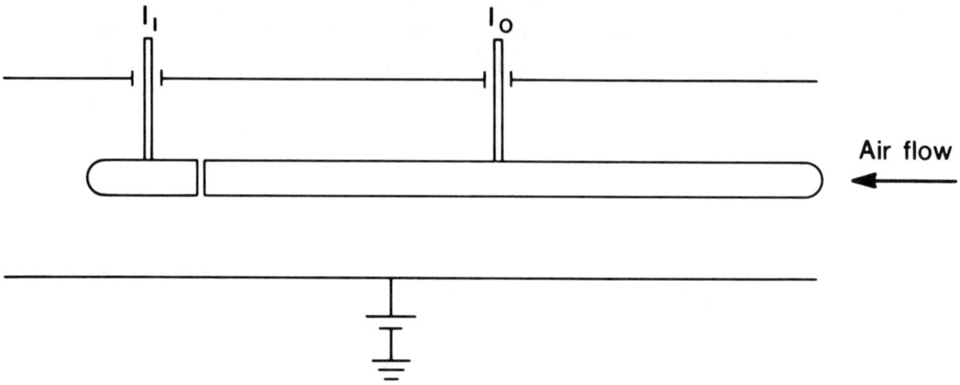

FIGURE 15. Divided collector ion counter used to obtain mobility spectra of atmospheric air ions.

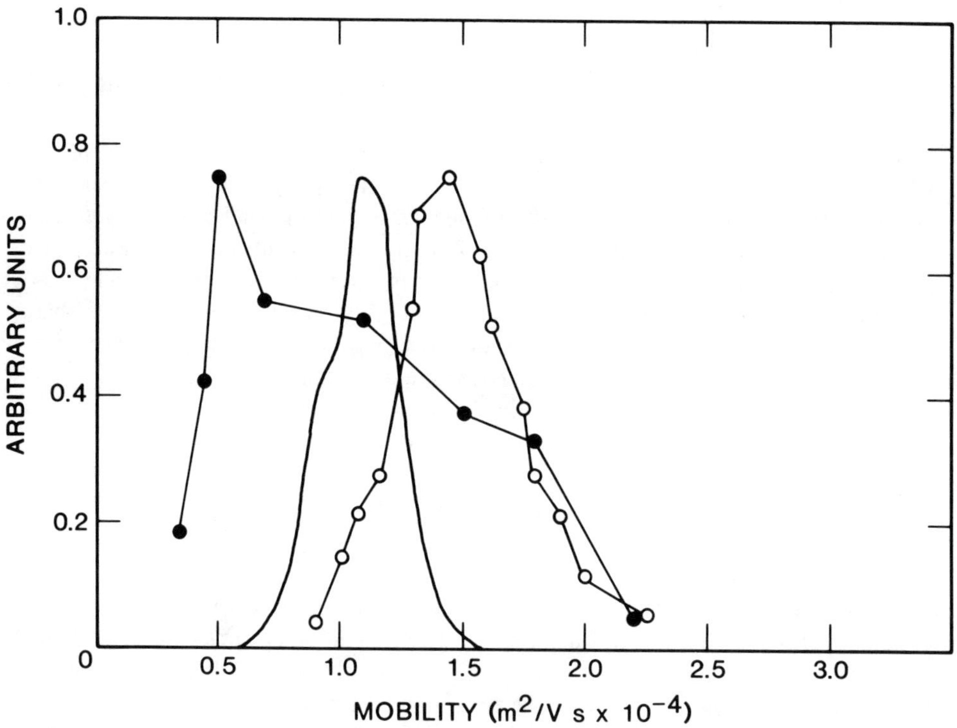

A

FIGURE 16. Mobility spectra for air ions determined from ion counter characteristics; (A) positive ions; (B) negative ions. (○ from Whipple, E. L., *J. Geophys. Res.*, 65, 3679, 1960; ● from Hoppel, W. A., *Pure Appl. Geophys.*, 81, 230, 1970; time-of-flight measurements from Huertas, M. L., *Aerosol Sci.*, 2, 145, 1971.)

number of applications, including the comparison of ions produced by corona discharge and radioactive sources,[58] and the measure of mobility of spray electrified liquids.[74] Modern devices used to provide information about aerosol size distributions are based on a merger of cylindrical ion counters and the air flow method.[67]

The use of air flow methods has provided low resolution information about the mobilities of atmospheric ions, both near the surface of the earth and in rocket and balloon flights into

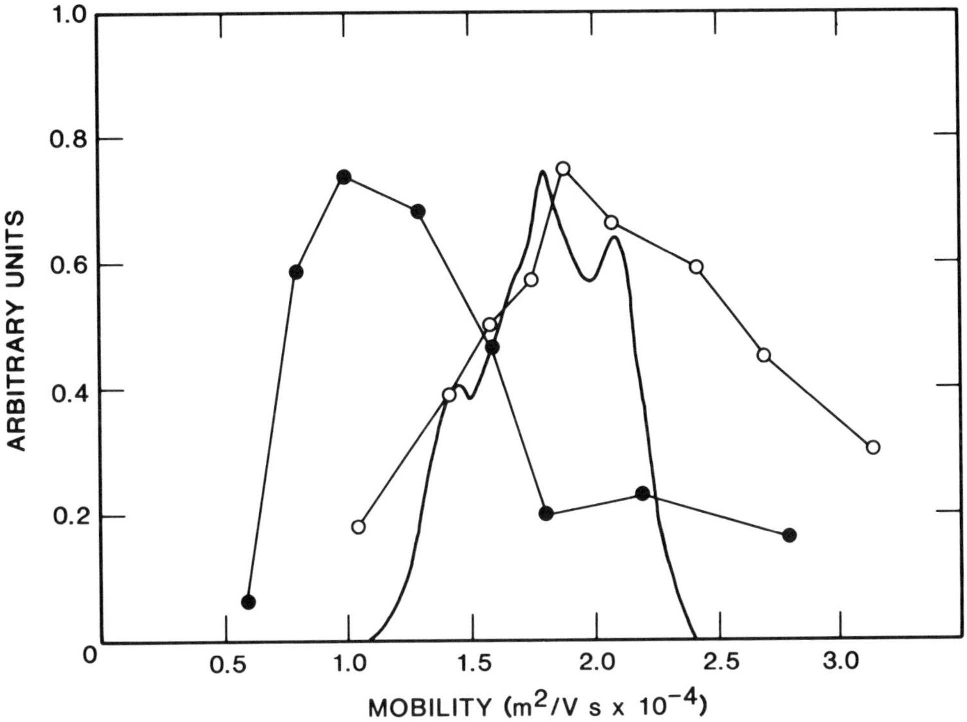

FIGURE 16B.

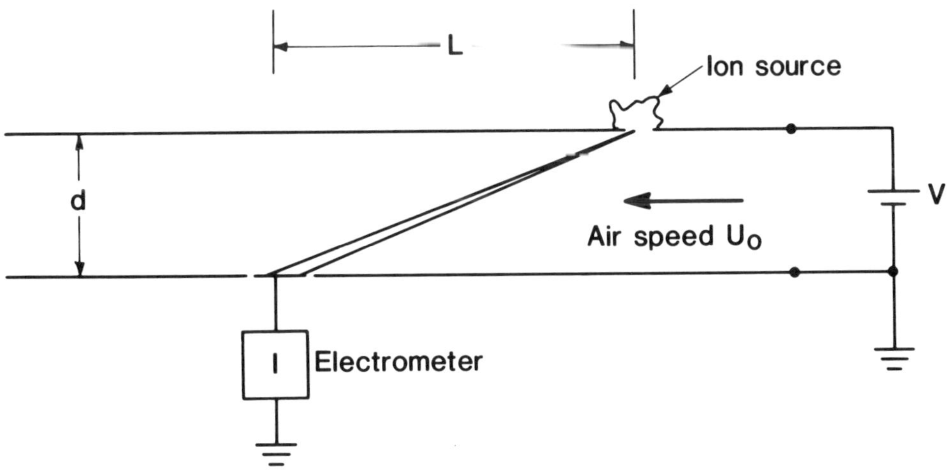

FIGURE 17. "Air Blast" method for measuring ion mobilities. Spectrum is obtained directly in this differential method.

the upper atmosphere.[75-78] More sophisticated methods for determining ion mobilities have been used in laboratory investigations at reduced pressures, and the most common method is based on time-of-flight techniques.[79] The conceptual features of a time-of-flight instrument are shown in Figure 18. Ions are produced in the source region and enter the drift region upon application of a control pulse to the control "gate". The ions move through the drift region, which may contain various background gases with which the drifting ions interact. After some drift time, determined by the ion mobility, drift region electric field, and length,

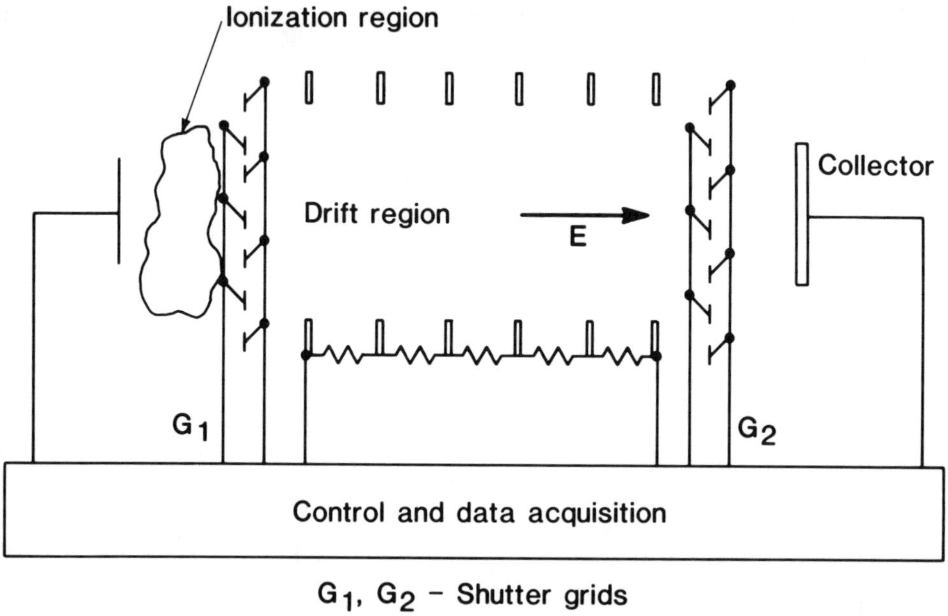

FIGURE 18. Time-of-flight apparatus for measuring ion mobilities (schematic).

and the ions pass through the second gate and arrive at the detector, producing a signal which is recorded. It is clear that by properly choosing the time relationship between pulses controlling the two gates, only ions in a given mobility range will reach the detector. By changing the delay time, a mobility spectrum is obtained.

Modern laboratory techniques have been devised to allow the study of ion processes in the laboratory, and a vast literature exists describing both the instruments used and the results of investigations. An example of a laboratory facility, which can be used to provide information about both the mobility and composition of ions produced at atmospheric pressures, is shown schematically in Figure 19. The facility combines a time-of-flight section with a mass analysis system. Ions which have mobilities as selected by the gates of the time-of-flight apparatus enter the mass analysis system. By using these devices in tandem, masses of ions having particular mobilities can be determined. Complete systems of the type shown represent a substantial investment and are intended for use in basic research. Detailed descriptions of examples of such systems are beyond the scope of this handbook but may be found in the literature.[80,81]

Of primary interest in the present context are studies of air ions done at atmospheric pressure. A review of selected results indicates that the presence of different trace-level gases in air can substantially modify the mature ion species found in experiments involving laboratory air.[18] This same review shows clearly that the continuous mobility spectra found in air using classical methods, such as those shown in Figure 16, are, in fact, made up of discrete groups of ions with different mobilities.

The use of these laboratory techniques in the field has been limited. The direct measurement of mobility spectra using drift tube or other techniques should be straightforward, but would require the development of suitable apparatus designed specifically for use in the field.

An apparatus designed for the measurement and analysis of atmospheric ions at ground level has been described and used for an initial set of measurements.[82,83] The apparatus uses special collection and focusing techniques to allow data acquisition even under the extremely low ion concentrations found in the troposphere (approximately one part in 10^{17}).

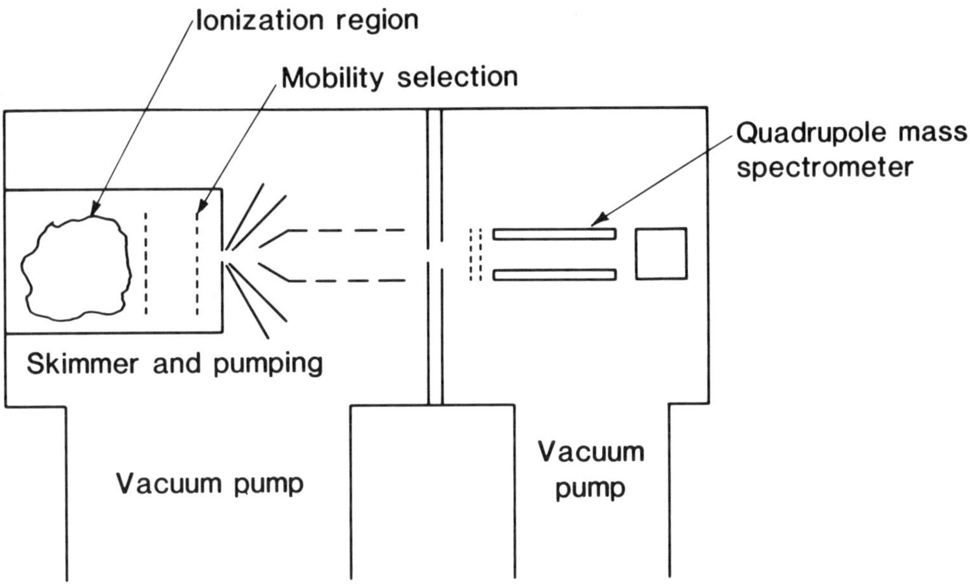

FIGURE 19. Representation of system for studying various aspects of ion mobilities and ion-molecule reactions. System includes mass spectrometer to allow mass analysis of mobility-selected ions.

IX. ERRORS ASSOCIATED WITH EXTERNAL ELECTRIC FIELDS AND OFF GROUND OPERATION OF INSTRUMENTS

Various sources of error associated with different measurements have been described in the previous sections. Errors due to external fields, losses in ducts, and internal space charge effects, are particular to aspiration devices and are discussed in more detail here.

A. Effects Due to External Fields

Of particular concern with many of the measurements described are errors introduced when the measurement is made above the ground plane. Operation of devices above the ground plane may take place for practical reasons, such as those associated with long-term measurements outdoors, or may be deliberately done to determine the variation of electrical quantities with altitude. Since the instrument is located in a region where there is an electric field, the ambient field will be perturbed. If the instrument can be maintained at the average potential of the region around it, then the perturbation will be minimized. If the instrument is grounded, which is the usual configuration, then the degree of perturbation of the electric field will depend on the shape and location of the instrument. If the perturbation of the field near the inlet is viewed as resulting from charge induced on the inlet, then it is clear that charge carriers will be attracted to or repelled from the device, depending on the polarity of the ions and the induced charge. This problem has been addressed with regard to conductivity measurements, and various authors have made mention of possible errors; there have been limited experimental investigations which show a general consistency with elementary theoretical analysis.[84-87]

In one series of measurements, bias voltages were placed on electrodes at the inlet of ducts leading to conductivity tubes to minimize external field effects.[88] These voltages were controlled by an electric field measuring device to take account of variations. Measurements made from airplanes or other devices, which may become charged, are subject to errors, and there has been literature discussion of possible effects.[89,90] Specific discussions of the effects of external fields on measurement of space charge density are found in Reference

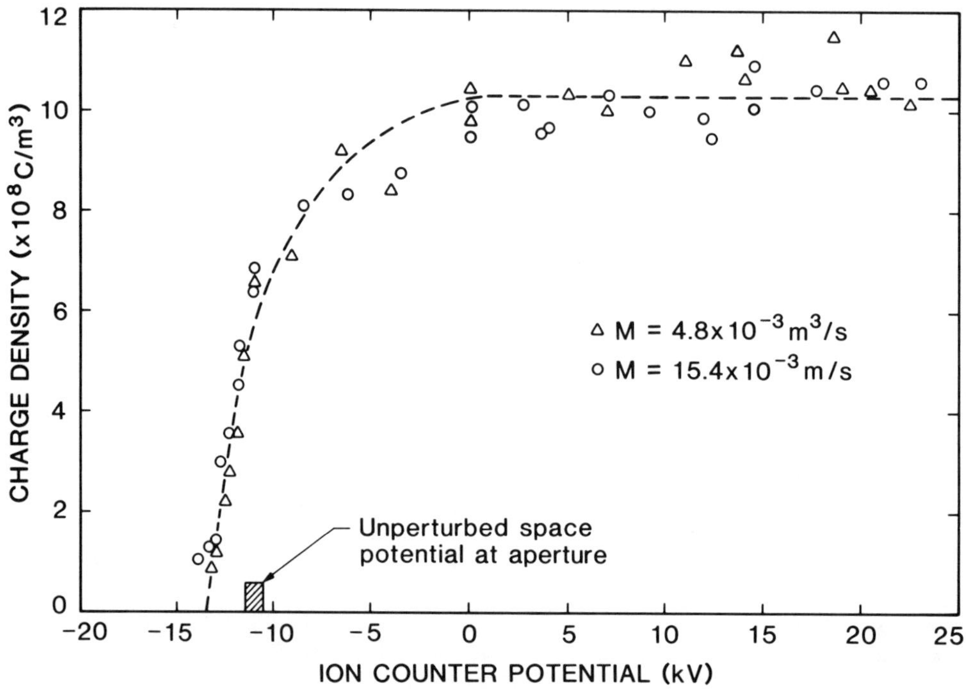

FIGURE 20. Above-ground ion density measurements at two air flow rates and as a function of ion counter potential. The unperturbed space potential at the height of the aperture, which is 2.54 cm in diameter, is indicated on the horizontal axis.

91. Schemes for maintaining devices at space potential have been described.[34,35] In order for a measuring system to be held at the average space potential around it, it must be connected to a high voltage supply which can respond to fluctuations in the potential. There appears to be little experimental effort described in the literature which addresses the magnitude of errors expected for various conditions.

A general conclusion from this earlier research suggests that for measurements of quantities related to ions attracted to the measuring device, the measurements were unperturbed, while significant errors occurred for those ions of the opposite sign which are repelled from the instrument. Laboratory experiments have been done using an ion counter located above the ground under a monopolar line.[92] In these experiments, unipolar ion densities were determined for different operating conditions. The results for one set of experiments are shown in Figure 20. Here the potential of the ion counter relative to ground was varied and the ion density calculated for each potential. The data shown were taken for two different flow rates through the instrument, and the scatter is representative of data taken over an extended period of time in a laboratory with poor control of environmental conditions. It is clear that there is little dependence on the potential of the counter when ions are attracted to the counter, which is consistent with the general conclusions stated above. The decrease of measured ion density as the potential becomes less attractive has no simple explanation, because of the complex geometry of the instrument, but appears to be associated with changes in the distribution of induced charges near the ion counter inlet.

When there are large (>1 kV/m) electric fields, as in biological exposure systems or near HVDC transmission lines, there is concern as to whether measurement errors can be substantial due to external fields, whether the measuring device is located above ground or not. In these situations, the ions originate from localized sources, such as radioactive sources or corona discharge, and are transported to the region of interest by electric fields, in contrast to the normal atmospheric situation where there is volume ionization. Here, the drift speeds of ions in the field may be substantially larger than those due to wind forces. The use of the parallel plate system (described previously) as a standard ion source has recently been investigated.[54] During the course of this study, comparisons were made of the ion density determined by using a parallel plate ion counter and those calculated from the physical characteristics of the parallel plate system. Where possible, corrections were made for duct losses. When this was done, there was agreement between the two methods for ion densities of approximately 6×10^5 ions/cm^3, external electric fields of 15.6 kV/m and average air speeds near the inlet from 1 to 6 m/s. These results are consistent with the discussions presented in earlier, classic papers[84-87] despite the large differences between the external electric fields used in the present study and those encountered in atmospheric electricity studies.

The problem of losses in ducts involves many variables, and only a limited number of investigations has been published.[89,90] Losses result predominantly from recombination, diffusion, turbulent transport, and self charge effects. The contribution from each of these, in turn, depends on other factors, including the physical dimensions of the duct, the characteristics of the air flow through the ductwork, and the nature of the space charge being measured. In general, if ductwork involves turns or is of any length, the individual experiment should estimate losses in the system being used and for the space charge conditions existing at the time of the measurement. This can be done by making related measurements in which the duct length is varied or bends are eliminated. Because of the self charge or ions, losses can be particularly large where ions of one polarity dominate. A recent series of papers describing the dispersal of ions provides guidance in estimating such losses and in placing bounds on the maximum space charge density that can exist for a given set of experimental conditions, but should not be used to replace experimental measurements.[95-96]

A general expression is given in Reference 95 for the space charge density after a time τ has elapsed, if the original density is ρ_o. This expression is

$$\rho = \frac{\rho_o}{1 + \rho_o \dfrac{k\tau}{\epsilon_o}} \tag{10}$$

where τ is the transit time of the ions from the source to the point of interest, k is the mobility, and ρ_o the original charge density. It is useful to estimate the loss in a duct using this equation. Figure 21 shows the predicted loss in ion density due to transit time effects. The ion density was assumed to be unipolar with initial values as indicated and the air speed at which the ions move in the duct was taken as 1 m/s, which is representative of air speeds associated with aspiration instruments. An ion densities become larger than 1×10^5/cm^3, the losses increase rapidly. It is more difficult to estimate losses due to diffusion because of uncertainties in transport mechanisms across boundary layers. In Reference 54 an estimate is made of losses due to diffusion to the walls of the duct transporting ions to an ion counter. A different approach using turbulent diffusion theory has been used to analyze a number of experimental results.[97]

Clearly, any ducts used in measurements should be made as short as possible and, as indicated above, experimental determination of the losses should be made whenever possible.

B. Space Charge Effects

Polarizing potentials are used to establish electric fields in the interior of conductivity tubes and ion counters. The presence of charge in the device will modify the applied field and may disturb the normal operation. Simple models to estimate the effects due to charge have been described in the literature.[73,97] These models are quite simple and ignore dynamics,

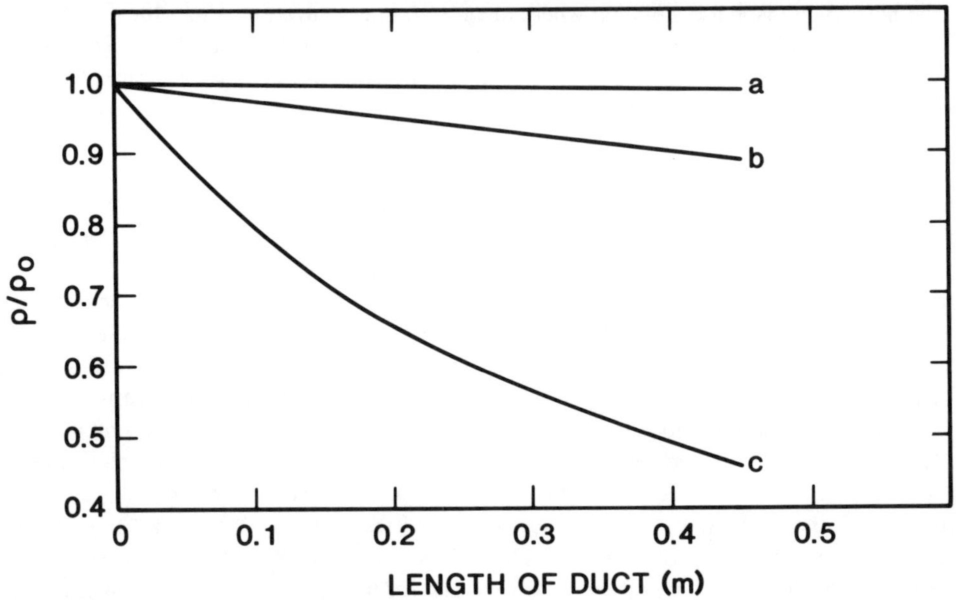

FIGURE 21. Estimated losses in a duct for an average speed of 1 m/s. Initial ion densities were taken as (a) 1 × 10^4/cm^3, (b) 1 × 10^5/cm^3 and (c) 1 × 10^6/cm^3. (Calculations based on theory in Kaune, W. T., Gillis, M. F., and Weigel, R. J., *J. Appl. Phys.*, 54, 6267, 1983.)

i.e., air flow and motion of charges. It is, nevertheless, worthwhile to present calculations based on these models to demonstrate the magnitude of the perturbation. In the references, Poisson's equation is solved for a cylindrical geometry, assuming that the space between the cylinders is filled with a uniform, immobile space charge. Figure 21 illustrates the results of model calculations for a cylindrical geometry having an inside radius of 0.75 cm, an outer radius of 5 cm, and an applied voltage of 10 V. It is assumed that the device is configured to collect positive ions and that the space charge is positive. For the example chosen, the electric field distribution is only slightly perturbed for space charge densities up to 1 × 10^5/cm^3, but for higher densities the changes are large. It is not clear what effect these perturbations have on the measurements themselves. For conductivity measurements, it would be necessary to adjust the flow rate or polarizing voltage to ensure that the measurement was being made on the linear portion of the I-V curve. If an ion counter measurement were being made, the critical mobility would be affected and it would be necessary to verify experimentally that the instrument was being operated in a saturated mode. The most serious effects would occur if mobility measurements were being made, since ions would be collected with no polarizing voltage and a normal I-V curve could not be generated. Suggested methods for overcoming these difficulties are found in Reference 98. A limited set of experiments has been described in which the operational characteristics of a Gerdien ion counter were determined for different unipolar space charge densities.[99] Within experimental uncertainty, no significant perturbations of ion counter operation were observed for ion densities up to 1 × 10^5 ion/cm^3. These results are consistent with the static charge calculations shown in Figure 22. A novel ion counter design has recently been described for use in situations where ions of one sign are predominate, but measurements of both unipolar charge densities are desired.[100]

X. APPLICATIONS OF MEASUREMENTS

A. HVDC Transmission Lines

HVDC transmission lines are, for practical design reasons, almost always in corona and,

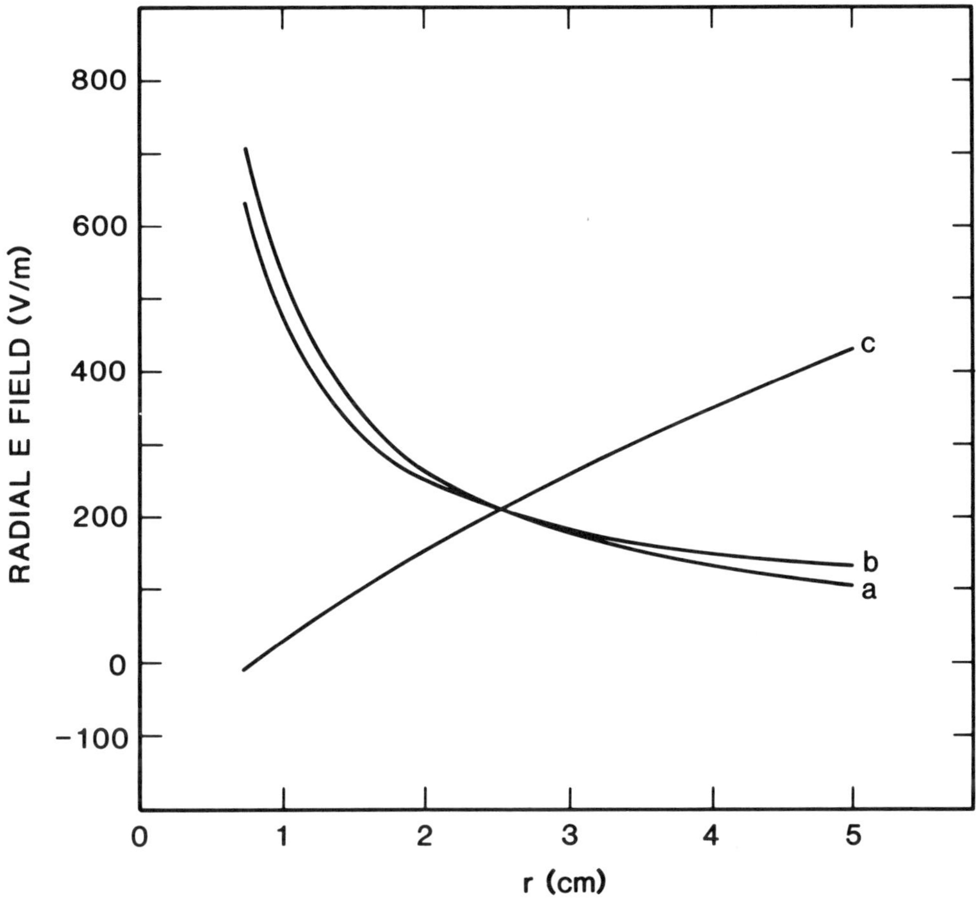

FIGURE 22. Effects of space charge on electric field in a long cylindrical geometry with inner radius of 0.75 cm and outer radius of 5 cm. Assumed space charge densities are (a) $1 \times 10^4/cm^3$, (b) $1 \times 10^5/cm^3$; (c) $1 \times 10^6/cm^3$. All dynamical effects have been ignored in these calculations. The voltage applied between the electrodes was taken as 10 V.

as a result, the electrical environment near these lines is significantly different from the ambient. Measurements of audible and radio noise, vertical current density, and electric field, are commonly made to characterize the corona performance of such lines and to offer engineering guidance for design.

Measurements of electric fields and ion densities near HVDC lines are of importance to those concerned with possible biological effects, since exposures for laboratory experiments are based on measurements made near actual lines. While the highest electric fields and small ion densities are found on or near the right of way, charge can be transported significant distances from the line by aerosols which become charged in the vicinity of the lines. This transport is mediated primarily by winds, so that a complete description of the physical phenomena involved must include meteorological data as well as a measure of the particulate density. In contrast to measurements made near the lines, those made at some distance involve much smaller signal levels (down to ambient), requiring more attention to instrumental details such as insulator integrity, electronic drift and sources of systematic error. Faraday cages are well suited for monitoring the transport of space charge away from HVDC transmission lines. Typical instrumentation used in studies of these lines is described in Chapter 20 and in the literature.[26]

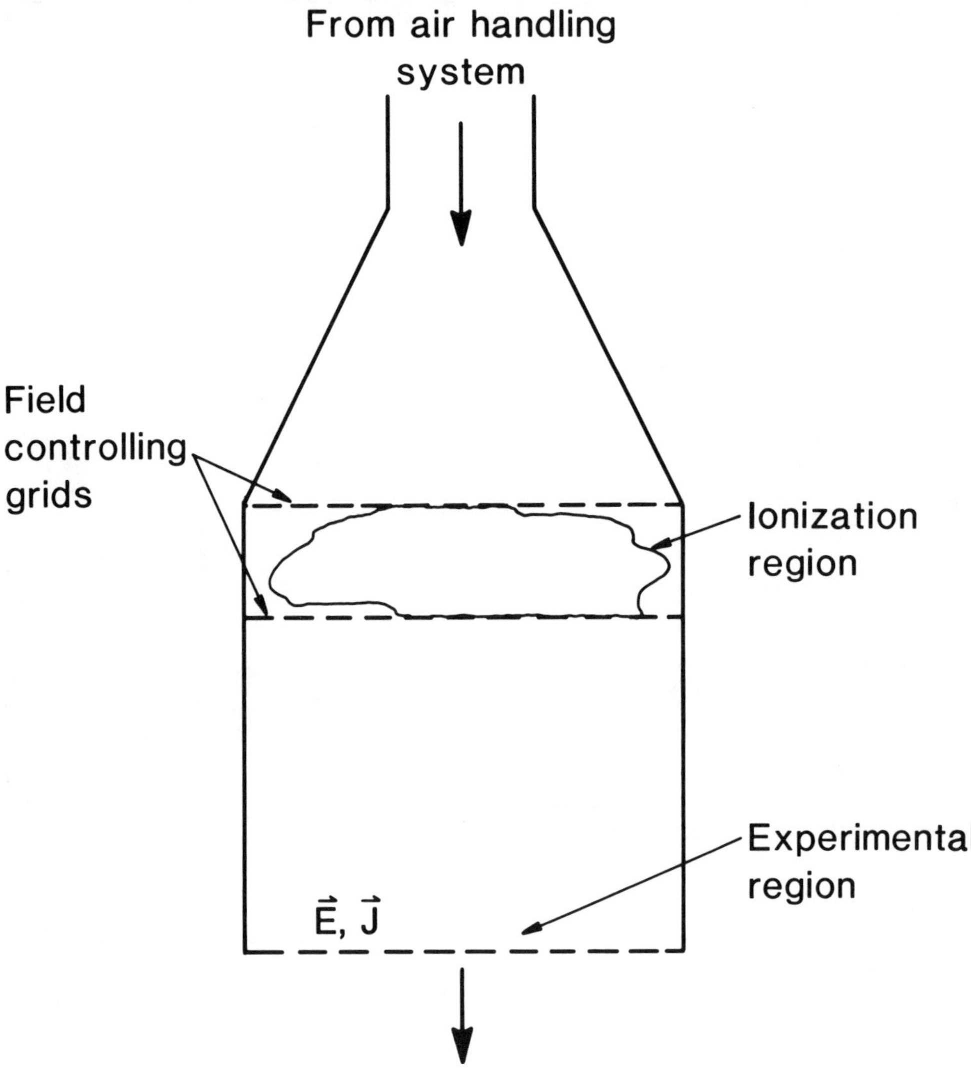

FIGURE 23. Schematic of exposure system for producing an environment containing electric field and space charge. System characteristics would be determined by measurements in or near the ground plane at the base of the experimental region. Details of animal caging, field grading assemblies, etc., are not shown.

B. Biological Exposure Systems

Biological exposure systems for studying the effects of dc fields and ions offer particular measurement challenges. It is important that the electrical characteristics of the exposure system be well known, since an integral part of the experiment is the dosimetry involved with determining exposure of the system under study. A simplified diagram of an exposure system is shown in Figure 23. This figure represents a composite of several real and proposed systems. There is a source of ions, a means of producing an electric field, an enclosure for the animals, and air flow for maintaining proper ventilation. Not indicated are various details of the caging or experimental apparatus.

An optimum chamber design would insure that the electric field and ion density would be uniform over the entire exposure region, remaining constant during the exposure period and be replicable from day to day. Examples of chamber designs are described elsewhere in this book. One feature of all the systems described is the presence of walls which both contain the animals and influence the electric fields and ion densities in the exposure region. Exposure systems offer particular challenges to making accurate measurements of electrical parameters because of their small size, limited access, and low air flow. It is fairly straightforward to determine the uniformity of electric field and vertical current density near the base of the exposure system by using an electric field probe and a small current sensing element. For best accuracy, the probe surface and current sensing element should be located in the ground plane. Consideration of measurement requirements during design and construction of the exposure facility should allow this to be done. If not, the systematic errors due to operation of the probes above the ground plane must be determined. This is a nontrivial task since the electric field near the enclosure walls may be nonuniform, and enhancement factors determined in uniform fields are not directly applicable.

Because of the presence of animals in the cages, it is probably not feasible to make direct measurements of electrical parameters at the time of exposure. Depending on construction of the exposure system, field and current density values may be determined in an area adjacent to the exposure region. Measurement of parameters which can be related to exposure is necessary if the experimental protocol is to be adequately specified.

Full-scale ion counters or filter devices operate with volumeteric air flows as large as 19 $\times$ 10^{-2} m^3/s (40 CFM). If such a device were used to measure ion densities in an exposure system, the disturbance of the ambient conditions would be significant. Such measurements are still useful if the extent of the perturbation can be estimated. Miniaturized devices, operating with reduced air flow, are described in the literature, but are not readily available.[50] A small ion counter using charged electrets to supply the polarizing potential has been studied, but is still in the developmental stage.[101] A discussion of ion characteristics and measurements from the viewpoint of the biologist in which certain types of ion sources and measurements related to them summarizes a number of the topics discussed above and points out several difficulties which can be encountered.[102]

C. Room Measurements

Some of the problems described above for making measurements in exposure systems are also in evidence when characterizing the electrical environment in other small volumes such as offices or laboratories. As noted, the air flow through an ion counter is significant and will certainly change the ambient conditions in a room, both by depleting the ion density and in altering the existing air flow. The introduction of a grounded object such as an ion counter will also modify the existing electric field distribution in the room.

Only a limited number of investigations involving rooms have been reported in the literature. A series of reports describing measurements of ions produced by various sources in a small room (volume of 16.5 m^3) has been published and is typified by References 97, 103, and 104. In these studies, the temporal behavior of ion densities and mobility distributions for large and small ions in the room was determined using ion counters. Measurements have also been reported describing the change of conductivity in sealed rooms.[105] An annotated bibliography on air and electrical field effects in biology contain references, published and unpublished, to measurements made in confined spaces.[106] Room measurements, using an HEPA filter to determine space charge in rooms, were used in a series of studies concerned with use of electrostatic air cleaners.[107] Although motivated for different reasons, in each of these cases measurements of ion densities and ion properties were of interest. Measurements of dc fields and ions along with other measurements designed to characterize the electrical environment in a closed room have recently been reported.[108]

ACKNOWLEDGMENTS

The work reported here, which was done at the National Bureau of Standards, was supported by the Electrical Energy Systems of the Department of Energy. Useful discussions with Dr. William Hoppel and Dr. Martin Misakian are gratefully acknowledged.

REFERENCES

The first two references present classical discussions of atmospheric electricity and are useful general references in addition to containing extensive bibliographies. Reference 3 is a conference proceedings which contains much useful information, while Reference 4 is an extensive discussion of aspiration methods. Specific citations are provided in later references for emphasis.

1. **Chalmers, J. A.,** *Atmospheric Electricity,* 2nd ed., Pergamon Press, Oxford, 1967.
2. **Israel, H.,** *Atmospheric Electricity,* Vol. 1 and 2 (transl.) U.S. Department of Commerce, Clearinghouse for Federal Scientific and Technical Information, Springfield, Va., 1973.
3. **Dolezalek, H. and Reiter, R., Eds.,** Electrical Processes in Atmospheres, *Proc. 5th Int. Conf. Atmos. Electr.,* Verlag, Darmstadt, 1977.
4. **Tammet, H. F.,** The Aspiration Method for the Determination of Atmospheric-Ion Spectra, (transl.) U.S. Department of Commerce, Clearinghouse for Federal Scientific and Technical Information, Springfield, Va., 1970.
5. **Bracken, T. D., Capon, A. S., and Montgomery, D. C.,** Ground level electric fields and ion currents on the Celilo - Sylmor ±400 kVDC intertie during fair weather, *IEEE Trans. Power Appar. Syst.,* PAS 97, 370, 1978.
6. **Comber, M. G. and Johnson, G. B.,** HVDC field and ion effects at project UHV: results of electric field and ion current measurements, *IEEE Trans. Power Appar. Syst.,* PAS 101, 1998, 1982.
7. **Vonnegut, B., Moore, C. B., Stout, G. E., Staggs, D. W., Brellock, J. W., and Bradley, W. E.,** Artificial modification of atmospheric space charge, *J. Geophys. Res.,* 67, 1073, 1962.
8. **Jones, C. D. and Griffiths, R. F.,** Full-scale experiments on dispersion around an isolated building using an ionized air tracer technique with very short averaging time, *Atmos. Environ.,* 18, 903, 1984.
9. **Penney, G. W. and Hewitt, G. W.,** Electrically charged dust in rooms, *AIEE Trans.,* 1949, 278.
10. **McCraty, R.,** Oper air ionization: winning the war against ESD, *Evaluation Eng.,* March, 78, 1983.
11. **Frazer, M. J. and Preache, M. M.,** Design, contruction and testing of a DC bioeffects test enclosure for small animals, Final Report, U.S. Department of Energy, Washington, D.C., RA 10157-1, 1980.
12. **Charry, J. M., Bailey, W. H., Shapiro, M. H., and Weiss, J. M.,** Ion exposure chambers for small animals, *Bioelectromagnetics,* 7, 1, 1986.
13. **Krueger, A. P., Staubbe, A. E., Yost, M. G., and Reed, E. J.,** Electric fields, small air ions and biological effects, Int. *J. Biometeorol.,* 22, 202, 1978.
14. **Weigel, R. J. and Kaune, W. T.,** System for exposing animals to DC electric fields and air ions: design considerations, presented at the Annu. Meet. Bioelectromagnetics Society, Washington, D.C., 1981.
15. **Lorrain, P. and Carson, D.,** *Electromagnetic Fields and Waves,* 2nd ed., W. H. Freeman, New York, 1970.
16. **Chalmers, J. A.,** *Atmospheric Electricity,* 2nd ed., Pergamon Press, Oxford, 1967, 23.
17. **Huertas, M. L. and Fontan, J.,** Evolution times of tropospheric positive ions, *Atmos. Environ.,* 9, 1018, 1975.
18. **Mohnen, V. A.,** Formation, nature and mobility of ions of atmospheric importance, *5th Intl. Conf. Atmos. Electric.,* Verlag, Darmstadt, 1977.
19. **Keesee, R. G. and Castleman, A. W., Jr.,** Ions and cluster ions: experimental studies and atmospheric observations, *J. Geophys. Res.,* 90, 5885, 1985.
20. **Dennis, R., Ed.,** *Handbook on Aerosols,* U.S. Department of Commerce, National Technical Information Service, Springfield, Va., 1977.
21. **Hoppel, W. A.,** Determination of the aerosol size distribution from the mobility distribution of the charged fraction of aerosols, *J. Aerosol Sci.,* 9, 41, 1978.
22. **Gunn, R.,** The secular increase of the world-wide fine particle pollution, *J. Atmos. Sci.,* 21, 168, 1964.
23. **Vosteen, R. E.,** D. C. electrostatic voltmeters and fieldmeters, Conf. Rec. 9th Annu. Meet. IEEE Indust. Appl. Soc., 1974.
24. **Mapelson, W. W. and Whitlock, W. S.,** Apparatus for the accurate and continuous measurement of the earth's electric field, *J. Atmos. Terrestrial Phys.,* 61, 1955.
25. **Kotter, F. R. and Misakian, M.,** *AC Transmission Line Field Measurements,* National Bureau of Standards, NBSIR 77-1311, National Technical Information Service, Report HCP/T-6010/E1, Springfield, Va., 1977.
26. **Comber, M. G., Kotter, R., and McKnight, R. H.,** Experimental evaluation of instruments for measuring DC transmission line experimental fields and ion currents, *IEEE Trans. Power Appar. Syst.* PAS 102, 3549, 1983.
27. **Maruvada, P. S., Dallaire, R. D., and Pedeault, R.,** Development of field-mill instruments for ground-level and above-ground electric field measurement under HVDC transmission lines, *IEEE Trans. Power Appar Syst.,* PAS 102, 738, 1983.

28. **Misakian, M.,** Generation and measurement of DC electric fields with space charge, *J. Appl. Phys.*, 52, 3135, 1981.
29. **Wildman, P. J. L.,** A device for measuring electric field in the presence of ionization, *J. Atmos. Terrestrial Phys.*, 27, 417, 1965.
30. **Sheahen, T. P.,** Model of response of an electric field mill operating during suborbital flights, *Rev. Sci. Inst.*, 45, 171, 1974.
31. **Waters, R. T.,** A Cylindrical Fluxmeter for Corona Studies, *J. Phys. E*, 5, 475, 1972.
32. **Horenstein, M. N.,** Peak sampled vibrating-reed for the measurement of electric fields in the presence of space charge, *Rev. Sci. Inst.*, 54, 591, 1983.
33. **Chubb, J. N.,** Developments in electrostatic fieldmeter instrumentation, *J. Electrostatics*, 14, 349, 1983.
34. **Kasemir, H. W.,** Calibration of atmospheric electric field meters and the determination of form factors at Kennedy Space Center, NOAA Technical Memorandum, ERL APCL-1, 1971.
35. **Smeddy, M. and Chalmers, J. A.,** The double field mill, *J. Atmos. Terrestrial Phys.*, 12, 206, 1958.
36. **Gathman, S.,** Guarded double field mill, *Rev. Sci. Inst.*, 39, 43, 1968.
37. **Hidaka, K. and Fijuta, H.,** A new method for electric field measurements in corona discharge using Pockels device, *J. Appl. Phys.*, 53, 5999, 1982.
38. **Chalmers, J. A.,** *Atmospheric Electricity*, 2nd ed., Pergamon Press, Oxford, 1967, 128.
39. **Lane-Smith, D. R.,** Review of instrumentation for atmospheric electricity, 5th Intl. Conf. Atmos. Electric., Verlag, Darmstadt, 1977.
40. **Hill, M. L. and Hoppel, W. A.,** Effects of velocity and other physical variables on the currents and potentials generated by radioactive collectors in electric field measurements, 5th Intl. Conf. Atmos. Electric., Verlag, Darmstadt, 1977.
41. **Crozier, W. D.,** Measuring atmospheric potential with passive antennas, *J. Geophys. Res.*, 68, 5173, 1963.
42. **Chalmers, J. A.,** The measurement of the vertical electric current in the atmosphere, *J. Atmos. Terrestrial Phys.*, 24, 297, 1962.
43. **Kasemir, H. W. and Ruhnke, L. H.,** Antenna problems of measurement of the air-earth current, in *Recent Advances in Atmospheric Electricity*, Proc. 2nd Conf. Atmos. Electr., Smith, L. G., Ed., Pergamon Press, New York, 1958.
44. **Hutchison, W. C. A.,** Atmospheric electric field charge compensation when measuring air-earth conduction and precipitation currents, *J. Atmos. Terrestrial Phys.*, 28, 823, 1966.
45. **Willett, J. C.,** The turbulent electrode effect as influenced by interfacial ion transfer, *J. Geophys. Res.*, 88, 8453, 1983.
46. **McKnight, R. H., Kotter, F. R., and Misakian, M.,** Measurement of ion current density at ground level in the vicinity of high voltage DC transmission lines, *IEEE Trans. Power Appar. Syst.*, PAS 102, 934, 1983.
47. **Misakian, M.,** Calibration of flat 60-Hz electric field probes, *J. Bioelectro. Soc.*, 5, 1984.
48. **Chalmers, J. A.,** *Atmospheric Electricity*, 2nd ed., Pergamon Press, Oxford, 1967, 196.
49. **Hoppel, W. A.,** Measurement of the mobility distribution of tropospheric ions, *Pure Appl. Geophys.*, 81, 230, 1970.
50. **Israel, H.,** *Atmospheric Electricity*, Vol. 1 and 2 (transl.) U.S. Department of Commerce, Clearinghouse for Federal Scientific and Technical Information, Springfield, Va., 1973, 131.
51. **Knoll, M., Eichmerer, J., and Schon, R. W.,** Properties, measurement and bioclimatic action of "small" multimolecular atmospheric ions, *Adv. Electr. Electron. Phys.*, 19, 177, 1964.
52. **Tammet, H. F.,** *The Aspiration Method for the Determination of Atmospheric Ion Spectra*, (transl.) U.S. Department of Commerce, Clearinghouse for Federal Scientific and Technical Information, Springfield, Va., 1970, 41.
53. **Smith, L. G.,** On the calibration of conductivity meters, *Rev. Sci. Inst.*, 24, 998, 1953.
54. **Misakian, M., McKnight, R. H., and Fenimore, C.,** Calibration of Aspirator-Type Ion Counters and Measurement of Unipolar Charge Densities, *J. Appl. Phys.*, 61, 1276, 1987.
55. **Vonnegut, B. and Moore, C. B.,** A Study of Techniques for Measuring the Concentration of Space Charge in the Lower Atmosphere, Final Report, Contract AF 191604, 1958.
56. **Anderson, R. V.,** Absolute measurements of atmospheric charge density, *J. Geophys. Res.*, 71, 5809, 1966.
57. **Asaro, K.,** Electrostatic potential and field in a cylindrical tank containing charged liquid, *Proc. IEE*, 124, 1277, 1977.
58. **Johnson, G. B. and Zaffanella, L. E.,** Techniques for measurements of the electrical environment created by HVDC transmission lines, Proc. *4th Int. Symp. on High-Voltage Engineering*, 1983, to be published.
59. **Obolensky, W. H.,** Über elektrische Ladungen in der Atmosphare, *Ann. Phys. Lpz.*, 77, 644, 1925.
60. **Bent, R. B.,** The Testing of Apparatus for Ground Fair Weather Space Charge Measurements, *J. Atmos. Terrestrial Phys.*, 26, 313, 1964.

61. **Moore, C. B., Vonnegut, B., and Mallabon, F. J.,** Airborne filters for the measurement of atmospheric space charge, *J. Geophys. Res.*, 66, 3219, 1961.
62. **Paltridge, G. W.,** The retention of atmospheric radioactivity by fibrous filters, *J. Geophys. Res.*, 72, 1269, 1967.
63. **McKnight, R. H. and Kotter, F. R.,** A facility to produce uniform space charge for evaluating ion measuring instruments, IEEE Trans. Power Appar. Syst., PAS 102, 2349, 1983.
64. **McKnight, R. H.,** The measurement of net space charge density using air filtration methods, *IEEE Trans. Power Eng. Appar.*, in press.
65. **Chalmers, J. A.,** *Atmospheric Electricity*, 2nd ed., Pergamon Press, Oxford, 1967, 196.
66. **Beckett, J. C.,** Air ion measurement, presented at the Int. Conf. on Ionization of the Air, Am. Inst. of Medical Climatology, Philadelphia, 1961.
67. **Whitby, K. T. and Clark, W. E.,** Electric aerosol particle counting and size distribution measuring system for the 0.015 to 1 size range, *Tellus*, 18, 573, 1966.
68. **McKnight, R. H., Kotter, F. R., Misakian, M., and Hagler, J. N.,** 1981 Annual Report: Electric and Magnetic Field Measurements, National Bureau of Standards, NBSIR 82-2527, Washington, D.C., 1982, 13.
69. **Antonevich, J. N. and Blitshteyn, M.,** Measuring effectiveness of air ionizers, in Proc. Electr. Overstress/Electrostatic Discharge Symp., Las Vegas, 1983.
70. **Israel, H.,** *Atmospheric Electricity*, Vol. 1 and 2 (transl.) U.S. Department of Commerce, Clearinghouse for Federal Secientific and Technical Information, Springfield, Va., 1973, 87.
71. **Whipple, E. L.,** An improved technique for obtaining atmospheric ion mobility distributions, *J. Geophys. Res.*, 65, 3679, 1960.
72. **Huertas, M. L., Marty, A. M., Fontan, J., Alet, I., and Duffa, G.,** Measurement of mobility and mass of atmospheric ions, *Aerosol Sci.*, 2, 145, 1971.
73. **Erickson, H. A.,** The charge of mobility of the positive ion in air with age, *Phys. Rev.*, 17, 400, 1921.
74. **Chapman, S.,** Carrier mobility spectra of spray liquified liquids, Phys. Rev., 52, 184, 1937.
75. **Conley, T. D.,** Mesospheric positive ion concentrations, mobilities and loss rates obtained from rocket-borne gerdien condenser measurements, *Rad. Sci.*, 9, 575, 1974.
76. **Widdel, H. U., Rose, G., and Borders, R.,** Experimental results on the variation of electric conductivity and ion mobility in the mesosphere, *J. Geophys. Res.*, 81, 6217, 1976.
77. **Meyerott, R. E., Reagan, J. B., and Joiner, R. G.,** The mobility and concentration of ions and the ionic conductivity in the lower stratosphere, *J. Geophys. Res.*, 85, 1273, 1980.
78. **Rosen, J. M., Hofmann, D. J., Grugel, W., Berkinski, J., Miehnowski, S., Morta, J., Ogawa, T., and Olson, D.,** Results of an international workshop on atmospheric electrical measurements, *J. Geophys. Res.*, 87, 1219, 1982.
79. **McDaniel, E. W. and Mason, E. A.,** *The Mobility and Diffusion of Ions in Gases*, John Wiley & Sons, New York, 1973.
80. **Shanin, M. M.,** Mass-spectrometric studies of corona discharges in air at atmospheric pressure, *J. Chem. Phys.*, 45, 2600, 1966.
81. **Good, A., Durden, D. A., and Kebarle, P.,** Ion-molecule reactions in pure nitrogen and nitrogen containing traces of water at total pressures 0.5-4 torr. kinetics of clustering reactions forming $H^+(H_2O)n$, *J. Chem. Phys.*, 52, 212, 1970.
82. **Eisele, F. I.,** Direct Tropospheric Ion Sampling and Mass Identification, *Int. J. Mass Spectrom. Ion Proc.*, 54, 119, 1983.
83. **Perkins, M. D. and Eisele, F. L.,** First mass spectrometric measurements of atmospheric ions at ground level, *J. Geophys. Res.*, 89, 9649, 1984.
84. **Swann, W. F. G.,** On certain matters relating to the theory of atmospheric-electric measurements, *Terr. Magn. Atmos. Electr.*, 19, 205, 1914.
85. **Hewlett, L. W.,** Investigation of certain causes responsible for uncertainty in the measurement of atmospheric conductivity by the Gerdien conductivity apparatus, Terrestrial Magn. Atmos. Electr., 19, 219, 1914.
86. **Swann, W. F. G.,** The theory of the action of the earth's potential gradient in measurements with the Ebert ion-counter, *Phys. Rev.*, 21, 449, 1923.
87. **Norinder, H.,** Investigations with regard to the induced charge from electrical fields on the Ebert ion counter, *Arkiv. Mat., Astron. Fysik*, 15, 1, 1920.
88. **Hegazi, K. A. and Chalmers, J. A.,** Measurements of atmospheric electric conductivity near the ground, *J. Atmos. Terrestrial Phys.*, 28, 327, 1966.
89. **Coroniti, S. C., Parziale, A. J., Callahan, R. C., and Patten, R.,** Effect of aircraft charge on airborne conductivity measurements, *J. Geophys. Res.*, 57, 197, 1952.
90. **Kraakevik, J. H.,** The measurement of atmospheric conductivity, *J. Geophys. Res.*, 63, 161, 1958.
91. **Bent, R. B. and Hutchinson, W. C. A.,** Electric space charge measurements and the electrode effect within the height of a 21 m mast, *J. Atmos. Terrestrial Phys.*, 28, 53, 1966.

92. **McKnight, R. H. and Fulcomer, P. M.,** Operation of ion counters near high voltage DC transmission lines, Proc. 4th Int. Symp. on High Voltage Engineering, National Technical University, Athens, 1983.
93. **Cobb, W. E.,** Ion losses in the Gerdien condenser intake system, *J. Appl. Met.*, 7, 456, 1968.
94. **Faster, W. W.,** Deposition of unipolar charged aerosol particles by mutual repulsion, *Br. J. Appl. Phys.*, 10, 206, 1959.
95. **Kaune, W. T., Gillis, M. F., and Weigel, R. J.,** Techniques for estimating space charge densities in systems containing air ions, *J. Appl. Phys.*, 54, 6267, 1983.
96. **Kaune, W. T., Gillis, M. F., and Weigel, R. J.,** Analysis of air ions in biological exposure systems, near HVDC electric power transmission lines, in rooms containing ion generators, and near exposed humans and animals, *J. Appl. Phys.*, 54, 6274, 1983.
97. **Siksna, R. and Lindsay, R.,** Air ions produced by a tritium-ion generator. II. Measurements of ions in a room, Mobility, *Arkiv. Geofysik*, 3, 141, 1959.
98. **Tammet, H. F.,** The Aspiration Method for the Determination of Atmospheric Ion Spectra, (transl.) U.S. Department of Commerce, Clearinghouse for Federal Scientific and Technical Information, Springfield, Va., 1970, 50.
99. **Allen, N. L. and Teisseyre, Y.,** Characteristics of the Gerdien ion counter in the presence of a significant unipolar space charge, *J. Phys. D: Appl. Phys.*, 16, 2485, 1983.
100. **Moody, N. F.,** Design and construction of an improved, portable air ion counter, *Int. J. Biometerol.*, 28, 169, 1984.
101. **Brylawski, R. E.,** Preliminary Investigations into the Use of Electrets in Measurement of Air Ion Concentrations, Master of Science thesis, University of Wisconsin-Madison, 1982.
102. **Pethig, R.,** Survey of the methodology for evaluating negative air ions: relevance to biological studies, *Methods in Enzymo.*, 105, 238, 1984.
103. **Norinder, H. and Siksina, R.,** Ions produced in the air at atmospheric pressure by ultra-violet light from a quartz-mercury ore, *Arkiv. Fysik*, 5, 471, 1952.
104. **Siksna, R.,** Mobility spectra of ions formed in a room by negative corona discharge, *Arkiv. Fysik*, 5, 545, 1952.
105. **Smith, L. G. and Schilling, G. F.,** The variation of electrical conductivity within sealed rooms, *J. Atmos. Terrestrial Phys.*, 4, 314, 1954.
106. **King, G. W. K., Ed.,** Ionization of the air and electrical effects, in *Biology-Bibliography of Published References*, 7th ed., G. W. K. King Associates, 1979.
107. **Penney, G. W., and Ziesse, N. G.,** Soiling of surfaces by fine particles, *ASHRAE Trans.*, 74, VI.3.1, 1968.
108. **Reiter, R.,** On the presence and generation of AC and DC electric fields and small ions in closed rooms as a function of building materials, utilization, and electrical installation, *J. Geophys. Res.*, 90, 5936, 1984.

Chapter 4

TRANSIT-TIME METHOD FOR ESTIMATING AIR-ION CONCENTRATIONS

William T. Kaune

TABLE OF CONTENTS

I. INTRODUCTION

A small fraction of the molecules and particles in atmospheric air are electrically charged.[1] Considerable research has been performed to determine if changes in concentrations of these air ions can affect living organisms.[2-5] Recent measurements have shown that air-ion concentrations are significantly increased in the vicinity of high-voltage direct-current (HVdc) electric-power transmission lines.[6-9] Concerns about possible health effects to humans or animals working or living near these lines have resulted in a much more critical scrutiny of past research results and the initiation of new projects to determine whether exposure to air ions produces biological effects.

The predominant mechanisms for the natural production of ionized air molecules are cosmic rays and radioactivity.[1,10] There are, in addition, a number of man-made sources of air ions. Ions, when first created, are molecular in size and are called small air ions. They can conveniently be classified by their mobilities, that is, the velocities imparted to them by a unit electrical field. Typical mobilities for small ions are in the range 1 to 2×10^{-4} $m^2/V\text{-}s$.[1,10] Small ions may collide with and attach to larger particles in the air to produce charged aerosol particles (sometimes referred to as large air ions). These ions are characterized by considerably smaller mobilities; typical values are in the range 0.03 to 1.0×10^{-6} $m^2/V\text{-}s$.[1,10] Aerosol particles may be multiply charged, whereas small ions are always singly charged.

Accurate calculations of air-ion concentrations in most systems are quite difficult if the air-ion space-charge densities are large enough to contribute significantly to the overall electric fields.[11,14] This chapter describes a relatively simple method of analysis which can be used to estimate air-ion concentrations for a number of interesting situations. The main application of the method is to systems containing small air ions, but it may also have use in systems containing charged aerosols.

The key result of the method developed in this paper is an equation relating the ion concentration at a particular point, P, in space to the ion concentration at the ion source and the time required for the ion to travel from its source to P. This time, called the transit time, can in many cases be estimated. In those cases where it cannot, an alternative approach is developed in which a lower bound on the transit time is computed to set an upper bound on the space-charge density at a given point P.

This method, the transit-time method, has been developed in considerable detail and generality in two published papers.[15,16] Here, the subject will be presented in a less general and, it is hoped, more accessible manner. The level of presentation will, however, be sufficient to enable the reader to understand many of the applications of the method which are of current interest.

II. THEORETICAL METHODS

A. Transit-Time Method

Consider a system (Figure 1) containing a source of unipolar ions, all of which are characterized by a single mobility k. This model is a good approximation to real systems containing unipolar small air ions because the mobilities of small ions lie in a fairly narrow range.[10] The velocity $\overline{v}$ of an ion in this system is

$$\overline{v} = k\overline{E} + \overline{w}, \tag{1}$$

where k is mobility, $\overline{E}$ is the total electric field (produced by the ions themselves and by any other objects in the system which are electrically charged), and $\overline{w}$ is wind velocity. Following Kaune et al.,[15] diffusional contributions have been neglected in this expression.

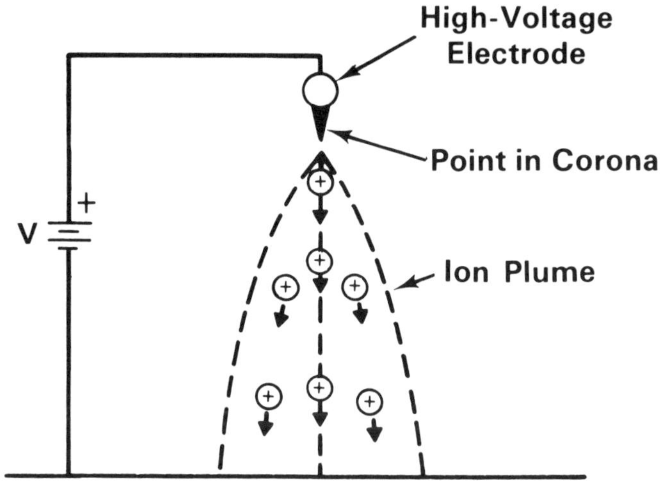

FIGURE 1. Corona discharge at a point electrode produces positive air ions. The electric field between the point and ground causes an ion plume to form. The diameter of the plume increases near ground because of the electrostatic repulsion between positive ions.

The current density $\bar{J}$ resulting from the motion of these (unipolar) air ions is

$$\bar{J} = \rho\bar{v} = \rho(k\bar{E} + \bar{w}), \tag{2}$$

where ρ, the space-charge density, is a function of position, $\bar{r}$, and time, t. The ion quantities $\bar{J}$, $\bar{E}$, and ρ are also related by the following two equations from electromagnetic theory:

$$\bar{\nabla}\cdot\bar{J} + \frac{\partial\rho}{\partial t} = 0 \text{ (Continuity equation)}, \tag{3}$$

$$\bar{\nabla}\cdot\bar{E} = \rho/\epsilon_o \qquad \text{(Guass' law)} \tag{4}$$

where t is time, $\bar{\nabla}$ is the gradient operator, and ϵ_o is the permittivity of air (8.85×10^{-12} F/m).

Equations 2 to 4 are difficult to solve in all but the simplest geometries. However, the following simple transformation leads to an explicit solution for ρ at a point P in space in terms of a single scalar parameter, the transit time, τ, required for an ion to move from its source to P. Thus, consider an observer who is riding with a moving ion, and let τ be time as measured by this observer. The space-charge density at this observer's location is $\rho[\bar{r}(\tau),\tau]$, where $\bar{r}(\tau)$ is the position of the ion (and the observer). The time derivative of ρ is

$$\frac{d\rho}{d\tau} = \bar{\nabla}\rho\cdot\frac{d\bar{r}}{d\tau} + \frac{\partial\rho}{\partial\tau}$$

Equation 1 can be used to eliminate the ion's velocity, $d\bar{r}/d\tau$, with the result that

$$\frac{d\rho}{d\tau} = \bar{\nabla}\rho\cdot(k\bar{E} + \bar{w}) + \frac{\partial\rho}{\partial\tau} \tag{5}$$

Using vector identities,[17] the following equation is easily derived:

$$\overline{\nabla}\rho\cdot(k\overline{E} + \overline{w}) = \overline{\nabla}\cdot\rho(k\overline{E} + \overline{w}) - \rho\overline{\nabla}\cdot(k\overline{E} + \overline{w})$$
$$= \overline{\nabla}\cdot\overline{J} - \rho k\overline{\nabla}\cdot\overline{E} - \rho\overline{\nabla}\cdot\overline{w}$$

Placing this result in Equation 5 yields

$$\frac{d\rho}{d\tau} = \overline{\nabla}\cdot\overline{J} - \rho k\overline{\nabla}\cdot\overline{E} - \rho\overline{\nabla}\cdot\overline{w} + \frac{\partial\rho}{\partial\tau} \tag{6}$$

According to Equation 3, the first and last terms on the right-hand side of Equation 6 add to zero. Furthermore, $\overline{\nabla}\cdot\overline{w} = 0$ to a good approximation (i.e., incompressible flow) if $\overline{w}$ is small compared to sonic velocities (Mach number less than about 0.2).[18] Under these assumptions, and using Equation 4 to eliminate $\overline{\nabla}\cdot\overline{E}$, we get

$$\frac{d\rho}{d\tau} = -\frac{k\rho^2}{\epsilon_o} \tag{7}$$

which has the solution

$$\rho(\tau) = \frac{\rho_o}{1 + \rho_o\left(\frac{k\tau}{\epsilon_o}\right)} \tag{8}$$

where ρ_o is the space-charge density at $\tau = 0$. This result was derived for the zero-wind case by Townsend in 1898.[19] Equation 8 is valid for both laminar and turbulent flow that satisfy the incompressibility condition.

One conclusion, which immediately follows from Equation 7, is that the magnitude of the space-charge density *always* decreases as time increases, i.e.,

$$\frac{d|\rho|}{d\tau} < 0 \tag{9}$$

Equation 8 can be simplified in the space-charge-saturated limit.[20] To see this, consider a system consisting of two electrodes with a potential difference, V, maintained between them. First, assume that there is no space charge present. The two electrodes will then be charged in such a way that the line integral of the electric field between them will equal V. Next, assume that one electrode — the source electrode — begins to emit ions that flow to the other electrode. Since the space charge will produce an electric field, the field produced by the source electrode will have to decrease in order to keep constant the line integral of the total electric field between the two electrodes. The space-charge-saturated limit is achieved when the electric field produced by the source electrode is reduced to zero. In this limit, the electric field at the surface of the source electrode becomes zero, which, by Equation 2, means that $|\rho_o| \rightarrow \infty$. (It is assumed that any wind is tangent to the surface of the source electrode.) Equation 8 then simplifies to

$$\rho(\tau) = \frac{\epsilon_o}{k\tau} \tag{10}$$

It is difficult to determine ρ_o in nonsaturated systems. However, Equation 8 can still be used to set an upper bound on $|\rho|$ by letting $|\rho_o| \rightarrow \infty$. The result is

$$|\rho| \leq \frac{\epsilon_o}{|k|\tau} \tag{11}$$

The results given by Equations 8 to 11 can be generalized to systems containing ions with several mobilities under steady-state (i.e., $\partial\rho/\partial\tau = 0$) and zero-wind conditions.[15]

B. Upper Bound on Space-Charge Densities

In order for Equations 8, 10, and 11 to be useful, transit times must be estimated. In certain systems (for example, where the wind is the dominant force on an ion) this can be done. However, in many cases where electric-field forces are significant, τ can not be determined in any simple way. This section develops an alternative approach, valid for steady-state conditions, in which a lower bound on the transit time of an ion is derived and inserted in Equation 11 to yield an upper bound on the space-charge density at a given point in space.

Denote by C the path along which an ion travels, let the initial and terminal points of this path be, respectively, at the ion source and at the point where we wish to estimate ρ, let s measure arc length along C, and assume that $\overline{w}$ is constant. (Cases where $\overline{w}$ is not constant are treated in Kaune et al.[15]) The transit time for the ion to traverse C is

$$\tau = \int_C \frac{ds}{\hat{s}\cdot(k\overline{E} + \overline{w})} \tag{12}$$

where $\hat{s}$ is a unit vector tangent to C in the direction of the ion's motion and $k\overline{E} + \overline{w}$ is the ion's velocity. The goal is to select a function $\hat{s}\cdot\overline{E}$ that minimizes τ. The functions that we can try for $\hat{s}\cdot\overline{E}$ are limited because they must satisfy the following relationship:

$$V = \int_C \hat{s} \cdot \overline{E}\, ds \tag{13}$$

where V is the potential difference between the ion's source and destination points.

The problem of minimizing τ, as defined by Equations 12 and 13, can be solved using the calculus of variations,[15] with the results that

$$\hat{s} \cdot \overline{E} = \frac{V}{S} + \frac{\overline{w}}{k} \cdot \left(\frac{\overline{D}}{S} - \hat{s}\right) \tag{14}$$

and

$$\tau_{min} = \frac{1}{kV/S^2 + \overline{D}\cdot\hat{w}/S^2} \tag{15}$$

where S is the length of the path C and $\overline{D}$ is a vector connecting the initial and terminal points of C. This result is still not in a useful form, however, because the shape of C, and hence S, will in general be unknown. Since $D \leq S$, this problem can be bypassed by replacing S by D in Equation 15, in which case we get that

$$\tau_{min} \geq \frac{1}{kV/D^2 + \hat{w}\cdot\overline{D}/D} \tag{16}$$

Combining this result with Equation 11 leads to the following upper bound on the space-charge density at the end of C:

$$|\rho| \leqslant \epsilon_o \frac{|V|}{D^2} + \epsilon_o \frac{\hat{w} \cdot \bar{D}}{|k|D^2} \tag{17}$$

This equation, Equation 8, and Equation 11 are the key results in this chapter.

III. SELECTED APPLICATIONS

A. Analysis of Symmetrical Systems

Using the transit-time formalism, a general solution can be given for the electric fields and space-charge densities in certain highly symmetric systems. These solutions can, of course, be obtained in a direct manner, but the transit-time formulation offers some simplicity, especially for space-charge saturated systems (i.e., $|\rho_o| \to \infty$) where the fact that the system is saturated can be introduced naturally into the analysis at its very beginning. The systems where this is possible are those where there is enough symmetry so that the geometrical path followed by an ion and the tangential current density along this path can be determined by inspection. These systems include: (1) parallel plates; (2) coaxial cylinders; and (3) concentric spheres. The solution for the space-charge densities and electric fields in these systems are obtained in the following manner.

Let z measure distance along the trajectory of an ion and let the current density tangential to the ion's path be J(z). Then

$$\rho \frac{dz}{d\tau} = J(z), \tag{18}$$

where $dz/d\tau$ is the velocity of the ion. Equation 8 can be used to express ρ in terms of τ. [For the space-charge saturated case, Equation 10 would be used instead.] With this substitution Equation 18 can be integrated:

$$\tau^2 + \left(\frac{2\epsilon_o}{k\rho_o}\right)\tau - \left(\frac{2\epsilon_o}{k}\right)\Gamma(z) = 0 \tag{19}$$

where

$$\Gamma(z) = \int_{z_o}^{z} \frac{dz'}{J(z')}, \tag{20}$$

$z = z_o$ is the location of the ion source and ρ_o is the space-charge concentration at $z = z_o$. Equation 19 can be solved for τ with the result that

$$1 + \rho_o\left(\frac{k\tau}{\epsilon_o}\right) = \left[1 + \left(\frac{2k\rho_o^2}{\epsilon_o}\right)\Gamma(z)\right]^{1/2}. \tag{21}$$

Using this result to eliminate τ in Equation 8 yields the following expression for the space-charge density:

$$\rho(z) = \rho_o\left[1 + \left(\frac{2k\rho_o^2}{\epsilon_o}\right)\Gamma(z)\right]^{-1/2}. \tag{22}$$

The electric field can be calculated from the expression $E = J/\rho k$ and Equation 21. The result is

$$E(z) = \frac{J(z)}{k\rho_o}\left[1 + \left(\frac{2k\rho_o^2}{\epsilon_o}\right)\Gamma(z)\right]^{1/2}. \tag{23}$$

In most experimental situations, the quantities which will be known are J(z) (through a measurement of the system's corona current) and the potential difference, V, between the electrodes of the system. V can be related to E(z) using Equation 13.

As an example, consider a system of two coaxial cylinders. Assume that the inner cylinder is a wire of radius a and that the potential difference between the two cylinders is sufficiently large so that the inner wire is in corona and the system is space-charge limited. A steady corona current, I_o, will be emitted per unit length from this ion source. By symmetry considerations it is evident that ions will move along radial paths from the inner to outer electrode and that the current density between the cylinders is $J(r) = I_o/(2\pi r)$, where r measures radial distance.

From Equation 20,

$$\Gamma(r) = \frac{\pi}{I_o}(r^2 - a^2), \tag{24}$$

where $z = r$ in this example. In the space-charge-saturated limit ($\rho_o \rightarrow \infty$), Equations 22 and 23 become

$$\rho(r) = \left(\frac{\epsilon_o I_o}{2\pi k}\right)^{1/2} \frac{1}{(r^2 - a^2)^{1/2}} \tag{25}$$

and

$$E(r) = \left(\frac{I_o}{2\pi\epsilon_o k}\right)^{1/2} \frac{(r^2 - a^2)^{1/2}}{r} \tag{26}$$

The potential difference between the electrodes is related to E as follows:

$$V = \int_a^b E(r)\, dr$$

$$= \left(\frac{I_o}{2\pi\epsilon_o k}\right)^{1/2} \left[(b^2 - a^2)^{1/2} - a \sec^{-1}\left(\frac{b}{a}\right)\right]. \tag{27}$$

where b is the radius of the outer electrode. This equation gives the relation between I_o and V in the space-charge-saturated limit.

The transit-time method (Equation 17) gives the following upper bound, ρ_{ub}, on the space-charge density produced at the surface of the outer electrode:

$$\rho_{ub} = \epsilon_o \frac{V}{(b - a)^2}. \tag{28}$$

It is interesting to see how close to this upper value the actual space-charge density at $r = b$ is. Equations 25 and 27 can be used to obtain an expression for $\rho(b)$ as a function of V. This value can then be divided by the value for ρ_{ub} in Equation 28 to give

$$\frac{\rho(b)}{\rho_{ub}} = \frac{(\eta - 1)^2}{\eta^2 - 1 - (\eta^2 - 1)^{1/2} \sec^{-1}(\eta)}, \tag{29}$$

where $\eta = b/a$. It can be shown that this ratio is always greater than 0.75 and that it approaches 0.75 for $b/a \rightarrow 1$ and 1.0 for $b/a \rightarrow \infty$. Thus, a system involving coaxial cylinders can produce space-charge densities arbitrarily close to the theoretical upper limit given by Equation 28.

B. Generation of Space Charge in a Room

There are currently available a number of devices intended to fill a room with unipolar (usually negative) small air ions. To our knowledge there is available no published way to relate the strength of the ion source (i.e., the ion current emitted by it) to the resulting minimum steady-state level of space charge in the room. Using the transit-time technique, a relationship between these two quantities can be obtained.

Ions emitted by a generator into a room will move toward and ultimately hit one of the room boundaries (e.g., a wall). Suppose that the goal is to produce everywhere in the room a space-charge density which satisfies $|\rho| \geqslant |\rho_{min}|$. Combining this inequality with Equation 11 yields $\epsilon_o/|k|\tau \geqslant |\rho_{min}|$, which means that the transit time τ for any ion to go from the generator to a wall cannot exceed $\epsilon_o/k\rho_{min}$. In order to have a stable ion concentration, the current I_o from the ion generator must therefore be sufficient to replace *every* ion in the room within this time period. That is, $|I_o\epsilon_o/k\rho_{min}| \geqslant |Q|$, where Q is the total space-charge in the room. Q can be eliminated by using the relationship $|Q| > \Omega|\rho_{min}|$, where Ω is the volume of the room, giving

$$|I_o| > \left|\frac{k\Omega\rho_{min}^2}{\epsilon_o}\right|, \tag{30}$$

or

$$|\rho_{min}| < \left|\frac{\epsilon_o\, I_o}{k\Omega}\right|^{1/2} \tag{31}$$

C. Decay of Space Charge in Ducts

The decay in concentration of a volume of air ions moving through a duct is of interest in several types of situations. For example, ducts may be used to carry air ions from an ion generator to an animal exposure chamber, or a series of ducts might be used to model the human respiratory tract. These kinds of systems can be analyzed using the transit-time formalism if air movement is the dominant influence on the motion of an ion through a duct.

Consider a duct of cross-sectional area A through which air is passing with a volumetric flow rate F. The average air velocity in the duct is $w = F/A$, and the time required for an ion to move a distance s in the duct is $s/w = sA/F$. If s is measured from the entrance of the duct, and the uniform space-charge density at the entrance is ρ_o, then the space-charge density a distance s downstream is

$$\rho(s) = \frac{\rho_o}{1 + \rho_o\left(\dfrac{kAs}{\epsilon_o\, F}\right)} = \frac{\epsilon_o\, F\rho_o}{\epsilon_o\, F + \rho_o\, kAs} \tag{32}$$

This equation has been derived by others[21-24] and has recently been used by McKnight et al.[25] to explain ion losses which occurred in a duct carrying air ions to an ion counter.

D. Evaluation of Published Space-Charge Data

Many papers have been published in which measured or calculated space-charge densities have been reported. The transit-time technique provides a method for checking the credibility of published data. The method will be illustrated by two examples.

1. Air Ion Exposure Chamber Used by Terry et al.

Terry et al.[26] reported a study in which maze learning was assessed in rats exposed to three concentrations of negative air ions. The experimental apparatus consisted of an enclosed environment chamber, measuring $91.5 \times 61 \times 61$ cm^3, within which rats were exposed to negative air ions. Four ion generators were placed in ports in the four walls of the chamber and negative ions, produced within these generators by radiation sources, were carried by air flow into the chamber. The minimum distance an ion traveled to reach the center of the chamber was 30 cm.

The authors reported that they measured a maximum exposure level of 7×10^7 ions/cm^3. If this measurement is correct, then Equation 11 shows that the maximum time an ion could have taken to travel 30 cm from its source to the center of the chamber was <5 ms. Since ions were carried into the chamber by air flow, and were opposed in this motion by the (very large) electric field generated by the ions already in the chamber, it may be concluded that air velocities in the chamber were well in excess of 60 m/s. It seems clear that this was not actually the case, so we must conclude that the authors' measurements were in error.

2. Air Ion Exposure Chamber Used by Krueger, Andriese, and Kotaka

Kruger, Andriese, and Kotaka,[27] and Krueger and Kotaka,[28] described an air-ion exposure chamber which consisted of a 23-cm-high, 12-cm-diameter plexiglass cylinder with a grounded metal floor on which mice resided, and a grounded metal roof on which was placed an ion generator whose electrical potential was maintained at +900 V. A grounded metal screen was located about 2 cm below the source. For transit-time calculations it is convenient to regard this screen as the ion source, in which case the voltage between the source and the grounded floor of the exposure cage was zero. The wind velocity between the source and cage floor was measured to be 0.28 m/s,[27] so the transit time for an ion to move from the roof to the floor was about 0.82 s. Using Equation 10, we estimate that the space-charge density in this system was about 8×10^{-8} C/m^3, which corresponds to an ion concentration of about 5×10^5 ions/cm^3. Measured values[28] were in the range 3.5 to 5.0×10^5 ions/cm^3 and are, therefore, consistent with theory.

E. HVdc Transmission Lines

High-voltage direct-current (HVdc) transmission lines offer advantages over alternating-current transmission lines for the long-distance transmission of electric power and when it is desirable to transfer power between two power systems which are not synchronized (i.e., not operating with exactly the same frequency and phase). HVdc lines are normally bipolar, that is, they have two bundles of conductors, which are referred to as the positive and negative poles. (A bundle consists of one or more closely spaced wires.) The positive and negative poles are operated with positive and negative voltages, respectively, relative to ground. These voltages are generally large enough so that the conductors operate in corona. Thus they are sources of small air ions, some of which are propogated to the ground under the combined influence of air movement and the electric fields generated by the line. A human or animal standing near a HVdc transmission line may thus be exposed to small air-ion concentrations above ambient levels. Upper bounds on this exposure can be estimated using the transit-time method.

Consider first a monopolar line consisting of a single pole (either positive or negative) located a distance h above ground (Figure 2). For zero-wind conditions, the maximum ground-level air-ion concentration would be expected directly under the conductor. An upper bound for the magnitude of the space-charge density there can be deduced directly from Equation 17. The result is

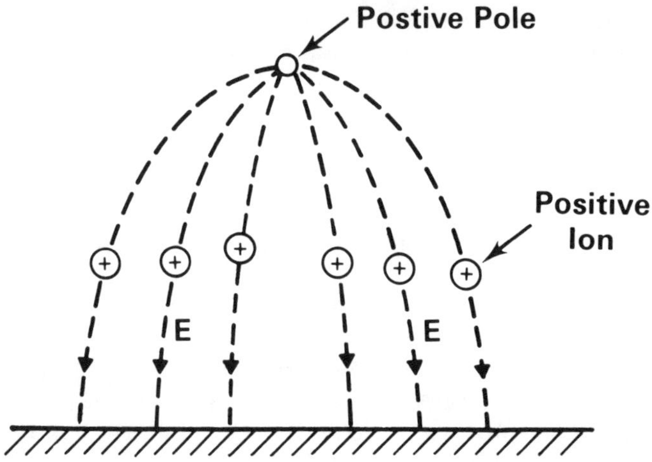

FIGURE 2. Cross section of a monopolar high-voltage direct-current transmission line. The positive pole may consist of two or more subconductors. Corona discharge at the surface of the pole creates air ions. Positive ions are carried to the ground because of the electric field produced by the line.

$$|\rho| \leq \epsilon_o \frac{|V|}{h^2}, \tag{33}$$

where V is the voltage of the pole.

Equation 33 can be tested by comparing it with maximum ground-level space-charge densities, ρ_{max}, measured under a scale-model monopolar transmission line.[29,30] For both positive and negative air ions, the experimental data are summarized by

$$\rho_{max} = (1.1 \times 10^{-11} \text{ C/Vm}) \frac{|V|}{h^2}. \tag{34}$$

This value is 25% higher than the upper bound expressed by Equation 33. It is, however, uncertain because the space-charge data upon which it is based were not directly measured but were instead calculated using the formula $\rho = J/(kE)$ and measured electric-field and current-density data. Thus, Equation 34 depends directly on the values used for the mobilities of positive and negative ions. The researchers who published Equation 34 used 1.15 and 1.50×10^{-4} m²/Vs, respectively, for these values.[29] Most publications, however, have used values closer to 1.4 and 1.7×10^{-4} m²/Vs.[20,31] It is interesting that if these latter values had been used in Equation 34, the constant on the right-hand side would have been 9.0×10^{-12}, which is very close to ϵ_o.

The extension of Equation 33 to bipolar HVdc transmission lines can be justified theoretically for zero-wind, steady-state conditions (and assuming the ground is not in corona)[16] and is supported by experimental data from scale-model lines which show that the ground-level ion concentration is maximum under monopolar lines.[29] Equation 33 can be tested for bipolar lines by comparing it to measured space-charge densities under actual operating and test HVdc transmission lines.[7,9,29,32] This comparison is made in Table 1 and shows that, within experimental uncertainties, space-charge densities derived from measurement are indeed bounded above by $\epsilon_o|V|/h^2$. It is interesting that full-scale HVdc lines, as well as scale models, can produce space-charge levels at times closely approaching the theoretical value given by Equation 33. This observation suggests that the simple expression in Equation 33 may be useful to those assessing the environmental impacts of currently operating HVdc transmission lines.

Table 1
SUMMARY OF MEDIAN SPACE-CHARGE DATA FROM OPERATING AND TEST HVdc TRANSMISSION LINES

	Height (m)	Pole spacing (m)	Number conductors/pole	Voltage (kV)	Weather	Space-charge density (nC/m^3): From measured data	Space-charge density (nC/m^3): Upper bound	Ref.
Nelson River	15.3	13	2	±450	Fair	10 −5	17 −17	9
Pacific Intertie	12	12	2	±400	Fair	8 −12	25 −25	7
Dalles Test Line	12	12	2	±533	Fair	21 −18	33 −33	31
Dalles Test Line	12	12	2	±500	Snowing	18 −7	31 −31	31
Project UHV	12	24	4	±800	Fair	38 −13	49 −49	28
Project UHV	12	24	4	±800	Raining	43 −32	49 −49	28
Project UHV	24	24	4	±800	Fair	8 −7	12 −12	28
Project UHV	24	24	4	±800	Raining	10 −8	12 −12	28
Project UHV	14	32	6	±1200	Fair	60 −3	54 −54	28
Project UHV	20	32	6	±1200	Fair	20 −4	27 −27	28

Note: Positive and negative space-charge (ρ) values are median ground-level values measured under the positive and negative poles, respectively. These values were calculated from measured median electric-field (E) and current-density (J) data using the equation $\rho = J/(kE)$. For this calculation the mobilities (k) of positive and negative small ions were assumed to be 1.4 and 1.7 $cm^2/V\text{-}s$, respectively. The transit-time method was used to calculate the upper bounds on space-charge density.

The analysis given in this section has been extended to windy conditions and to locations not directly under a pole of a HVdc transmission. This development is carried out in Kaune et al.[16]

IV. DISCUSSION

The transit-time method described in this paper offers a relatively simple means for estimating air-ion concentrations in many systems of interest. Several examples are given in the preceding section of this paper. The method also provides an alternative way of thinking about air ions which can be very useful in some cases. The following two paragraphs present one such example.

Suppose that it is desired to increase the air-ion concentration in a system where air ions are generated by a localized source and are carried to the target by air flow. On first consideration it would appear that the way to do this would be to increase the strength of the ion source (e.g., by adding radioactive material). However, the transit-time formalism predicts that this change will not necessarily accomplish the desired goal.

In most cases where ions have a localized source, it will be true that $\rho << \rho_o$ at any significant distance from the source, in which case Equation 8 simplifies to $\rho = \epsilon_o/(k\tau)$. This equation shows that ρ can be increased *only* by decreasing τ. In some systems, τ will be decreased by increasing ρ_o because an increase in ρ_o will cause an increase in the electric fields between the source and the target. However, in systems like our example where the dominant influences on the motion of an ion are air flow and/or electric fields produced by sources other than the ion cloud itself, τ will not be affected by an increase in ρ_o and there will, therefore, be no corresponding increase in ρ.

ACKNOWLEDGMENTS

The work described in this paper was supported by the Electric Power Research Institute. The author would like to acknowledge helpful conversations with Drs. Murlin F. Gillis, Richard J. Weigel, Ronald McKnight, and Martin Misakian.

REFERENCES

1. **Chalmers, J. A.,** *Atmospheric Electricity,* Pergamon Press, New York, 1967.
2. **Krueger, A. P.,** Air ions and physiological function, *J. Gen. Physiol.,* 45, 233, 1962.
3. **Wehner, A. P.,** Electro-aerosols, air ions and physical medicine, *Am. J. Phys. Med.,* 48, 119, 1969.
4. **Anderson, I.,** Effects of natural and artificially generated air ions on mammals, *Biometeorol.,* 16, 229, 1972.
5. **Sheppard, A. R.,** Biological effects of static electric fields and air ions in relation to dc power transmission, in *Proc. Workshop on Electrical and Biological Effects Related to HVdc Transmission,* Bracken, T. D., Ed., NTIS, Springfield, PNL-3121, 1979.
6. **Bracken, T. D., Capon, A. S., and Montgomery, D. V.,** Ground level electric fields and ion currents on the Celilo-Sylmar ±400 kV dc intertie during fair weather, *IEEE Trans. Power Appar. Syst.,* PAS-97, 370, 1978.
7. **Bracken, T. D.,** The HVdc transmission line environment, in *Proc. Workshop on Electrical and Biological Effects Related to HVdc Transmission,* Bracken, T. D., Ed., NTIS, Springfield, PNL-3121, 1979.
8. **Comber, M. G. and Johnson, G. B.,** HVdc field and ion effects research at Project UHV: results of electric field and ion current measurements, *IEEE Trans. Power Appar. Syst.,* PAS-101, 1998, 1982.
9. **Maruvada, S. P., Dallaire, R. D., Noris-Elye, O. C., Thio, C. V., and Goodman, J. S.,** Environmental effects of the Nelson River HVdc transmission lines — RI, AN, electric field, induced voltage, and ion current distribution tests, *IEEE Trans. Power Appar. Syst.,* PAS-101, 951, 1982.

10. **Israel, H.,** *Atmospheric Electricity,* Vol. 1, NTIS, Springfield, TT67-51394-1, 1971.
11. **Maruvada, S. P. and Janischewskyj, W.,** Analysis of corona losses on dc transmission lines. 1. Unipolar lines, *IEEE Trans. Power Appar. Syst.,* PAS-88, 718, 1969.
12. **Maruvada, S. P. and Janischewskyj, W.,** Electrostatic field of a system of parallel cylindrical conductors, *IEEE Trans. Power Appar. Syst.,* PAS-88, 1476, 1969.
13. **Hill, H. L., Capon, A. S., Ratz, O., Renner, P. E., and Schmidt, W. D.,** *Transmission Line Reference Book-HVdc to ± 600 kV,* Electric Power Research Institute, Palo Alto, Calif., 1977.
14. **Janischewskyj, W. and Gela, G.,** Finite element solution for electric fields of coronating dc transmission lines, *IEEE Trans. Power Appar. Syst.,* PAS-98, 1000, 1979.
15. **Kaune, W. T., Gillis, M. F., and Weigel, R. J.,** Technique for estimating space-charge densitites in systems containing in air ions, *J. Appl. Phys.,* 54, 6267, 1983.
16. **Kaune, W. T., Gillis, M. F., and Weigel, R. J.,** Analysis of air ions in biological exposure systems, near HVdc electric power transmission lines, in rooms containing ion generators, and near exposed humans and animals, *J. Appl. Phys.,* 54, 6274, 1983.
17. **Reitz, J. R. and Milford, F. J.,** *Foundations of Electromagnetic Theory,* Addison-Wesley, Reading, Mass., 1960.
18. **Whitaker, S.,** *Introduction to Fluid Mechanics,* Prentice-Hall, Englewood Cliffs, N.J., 1968.
19. **Townsend, J.,** Applications of diffusion to conducting gases, *Philos. Mag.,* 45, 471, 1898.
20. **Misakian, M.,** Generation and measurement of dc electric fields with space charge, *J. Appl. Phys.,* 52, 3135, 1981.
21. **Wilson, I. B.,** The deposition of charged particles in tubes, with reference to the retention of therapeutic aerosols in the human lung, *J. Colloid Sci.,* 2, 271, 1947.
22. **Foster, W. W.,** Deposition of unipolar charged aerosol particles by mutual repulsion, *Br. J. Appl. Phys.,* 10, 206, 1959.
23. **Whitby, K. T.,** Generator for producing high concentrations of small ions, *Rev. Sci. Instrum.,* 32, 1351, 1961.
24. **Whitby, K. T. and McFarland, A. R.,** The decay of unipolar small ions and homogeneous aerosols in closed spaces and flow systems, in *Proc. Int. Conf. on Ionization of the Air,* Franklin Institute, Philadelphia, 1961.
25. **Hebner, R. E., Ed.,** *Development of Power System Measurements-Quarterly Report,* Department of Energy, Division of Electrical Energy Systems, Washington, D.C.
26. **Terry, R. A., Harden, D. G., and Mayyasi, A. M.,** Effects of negative air ions, noise, sex and age on maze learning in rats, *Int. J. Biometeor.,* 13, 39, 1969.
27. **Krueger, A. P., Andriese, P. C., and Kotaka, S.,** Small air ions: their effect on blood levels of serotonin in terms of modern physical theory, *Int. J. Biometeor.,* 12, 225, 1968.
28. **Krueger, A. P. and Kotaka, S.,** The effects of air ions on brain levels of serotonin in mice, *Int. J. Biometeor.,* 13, 25, 1969.
29. **Comber, M. G., Nigbor, R. J., and Zaffanella, L. E.,** *HVdc Transmission Line Research,* Electric Power Research Institute, Palo Alto, Calif., EL-2419, 1982.
30. **Harvey, J. R. and Zaffanella, L. E.,** *DC Conductor Development,* Electric Power Research Institute, Palo Alto, Calif., EL-2257, 1982.
31. **Mohnen, V. A.,** Formation, nature, and mobility of ions of atmospheric importance, in *Proc. 5th Int. Conf. on Atmospheric Electricity,* Garmisch-Partenkirchen, Germany, 1974.
32. **Chartier, V. L., Bracken, T. D., and Stearns, R. D.,** Corona Phenomena, Electric Field, and Ion Measurements at the Dalles HVdc Test Line to Determine Feasibility of Uprating the dc Intertie, Bonneville Power Administration, Vancouver, ERJ-78-88, 1979.

Chapter 5

AIR ION EXPOSURE SYSTEMS FOR SMALL ANIMALS

Richard J. Weigel

TABLE OF CONTENTS

I. INTRODUCTION

Carefully controlled experimentation on the effects of air ions on animal physiology began in the Air Ion Research Laboratory, University of California at Berkeley, in the late 1950s. Since then, dozens of other laboratories have continued experimentation in which animals are exposed to concentrations of air ions above normal ambient levels. In each case, the researcher had to develop an exposure chamber which was suitable for the purposes of his experiment. As we shall see in this chapter, the design of an exposure system is a complex subject and its successful performance may be affected by many design and environmental factors. Many of the systems described in the literature have not demonstrated the ability to generate ions in the quantities claimed. In some cases, no evaluation of the exposure conditions or system is possible, as the system is not described in detail or even at all. This failing in many of the early experiments has led to criticism of the accuracy of some reported results.[1-3]

Part of the problem of poor exposure system documentation was a result of the lack of sophisticated instrumentation for the measurement of static electric fields and extremely low levels of air ions. With more sophisticated instrumentation, which only recently has become available, and with the realization that careful measurement and documentation of the exposure conditions are vital, current research is providing valuable data on the biological effects of air ions. This chapter will not be a comprehensive review of exposure systems used in the literature nor should it be considered as a manual for the construction of an air ion exposure system. Instead, I will discuss what I believe are some important considerations to observe as one designs and builds an exposure system. When developing new exposure systems, the points brought up in this chapter should be addressed in some manner.

As examples of good exposure system design, I will be discussing two recently completed exposure systems. One is a system developed at Rockefeller University by Charry and associates, and the other is the system designed at Battelle Pacific Northwest Laboratories by Weigel and Kaune. The basic difference between the two systems is that, while they both generate high concentrations of small ions, the Battelle system also generates a high strength static electric field while the smaller electric field in the Rockefeller system is a byproduct of the ion generation method. The Battelle system has been designed to simulate the electrical environment in the right-of-way of a high voltage dc transmission line, where high field strengths are found. Furthermore, the duration of exposure and number of animals exposed for the planned biological experiments are both much larger in the Battelle system, generating additional design criteria. These differences will become clear as we proceed.

II. EXPOSURE SYSTEM REQUIREMENTS

The design of any animal exposure system must generate uniform, stable, and known levels of the exposure agent and accommodate the specific requirements of the experiments to be conducted with the system. Systems designed for acute exposures will be less complex than those for chronic exposures. Food and water delivery and the ability to maintain a clean environment may not be needed in the acute system. Some factors which must be addressed in an exposure system design are considered below.

A. Environmental Factors

In general, small air ion exposure systems must provide the animals with a minimally stressful environment so as not to confound the results of the experiments. This is especially important when the expected effects of the exposure agent are small perturbations from physiological norms. Proper temperature and relative humidity, nonstressful sensory stimulation, food, and water must be provided. Food and water delivery may not be important for short-term exposures but for chronic exposures, they become major design factors. Primarily, in providing these supplies, one must be concerned that the water supply does not perturb the electric field nor shock the animal when drinking. Food delivery does not present as large a potential problem except that the food container could perturb the field by shielding if mounted above the ground plane.

The maintenance of a pure, low particulate air supply with stable temperature and relative humidity is extremely important whether the exposures are acute or chronic. The designer must be concerned that the ions being generated maintain their identities as small air ions. The collision of small ions with particulates results in the loss of ionic charge to the particle.[4] A large concentration of dust can greatly reduce the total concentration of small ions in the exposure system. Even when collisions with particulates do not occur, small air ions will undergo transitions and grow in size as they impact other molecules in air. The chemistry of small ions is beyond the scope of this chapter, other than to emphasize that ion size is significantly affected by relative humidity during the tenths of a second between creation of the ion and its arrival at the ground plane. This effect was illustrated by Johnson[5] who measured the mobility spectrum of laboratory generated small air ions of both polarities. As relative humidity increased, the spectrum showed an increased fraction of ions of lower mobilities, suggesting that ionic size had increased as more water molecules were bound to the charged ion. The longer the ion spent moving through the environment, the greater the shift of the spectrum towards lower mobilities.

What are the implications of this shift towards larger size for generating a "standard" small ion exposure? While no data are available on the significance of small shifts in ion size on their potential for biological effects, it remains prudent to include in the documentation of an air ion exposure system a description of factors which can affect the ion size. This would include the previously mentioned factors of relative humidity, temperature, particulate count, ion polarity, and also an estimation of the duration of the ions' path from the point of generation to the location where animal exposure occurs. This time estimate can be made using the transit time analysis discussed in the chapter by Kaune. Parameters which must be known to make this estimate are the dimensions of the exposure system, electric field strength, and the velocity of air in the cage.

B. Ion Generation Techniques

Air ions are generated whenever there is enough energy to strip an electron from an air molecule. Waterfalls, strong winds and cosmic rays are some of the naturally occurring processes which can produce air ions.[4] In the laboratory, radioactive decay and corona discharge are used to create equal numbers of positive and negative ions which are then separated by an electric field. Radioactive decay has frequently been used when the experimenter has chosen a design where ion generation occurs separate from the exposure chamber and air currents are used to transport them to the animals where only a small electric field exists.[6-8] Extreme care must be taken to insure that the desired concentration of ions does arrive at the exposure system. Without a significant electric field, the determination of space charge density requires the use of volumetric measurements of charge or extremely sensitive current-measuring devices (see chapter by McKnight). Guidance on the space charge density that can be expected can be obtained using the transit time analysis in the chapter by Kaune. Krueger et al. used a different method of generating and delivering air ions using radioactive decay.[9] Foils of tritium (a low energy beta emitter) impregnated zirconium were fixed to the top metal surface of the animal cages. A voltage placed on this surface generated the electric field which separated the two polarities of ions and delivered the selected polarity of ions to the animals. The concentration of ions reaching the animals was adjusted by varying the voltage on the disc.

At Battelle, we investigated a similar method of generating ions and identified certain

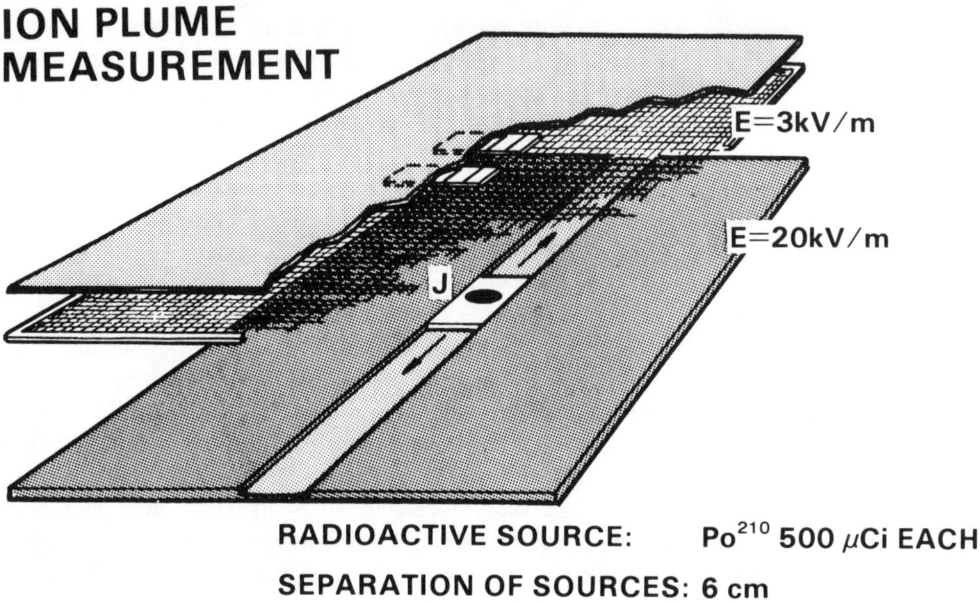

FIGURE 1. Method used to measure the distribution of ions generated by Polonium-210 sources in a parallel plate test system. Separation between top two plates is 8.5 cm; between the middle and ground plate it is 40 cm. Probe shown in slot in ground plane directly beneath the radioactive sources.

technical limitations which the exposure system designer should recognize. Our tests were conducted with a source made of five patches of polonium-210, an alpha emitter, each 0.5 × 0.6 cm, with a 0.5 cm gap between patches. Total activity of the five patches was 500 μCi. One of these sources was mounted to the underside of a 1 m^2 metal electrode and suspended 8.5 cm above a second electrode made of aluminum window screen. A grounded electrode was 40 cm beneath this electrode. An electric field meter and a current density probe (a 1 cm^2 metal sensor mounted in the ground plane and isolated from ground through the input of a high-impedance current-to-voltage converter) were used to scan the surface electric field and current density by sliding the instruments along a slot cut into the ground electrode. Separate dc power supplies were connected to the screen electrode and the electrode containing the polonium source. The electric field above the ground plane was established by the potential on the screen electrode. The number of ions which passed through the screen electrode was regulated by adjusting the voltage on the solid electrode. Various field strengths at the ground level were tested as well as various potential differences between the two electrodes. The tests demonstrated that the radioactive source generated a plume of ions with a width and current density within the plume that could be controlled by adjusting the electrode voltages. These two parameters, however, were not independent. Plumes with maximum ion densities of 400 nC/m^3 had widths less than 3 cm. The width of the plume is probably the major drawback to the use of discrete, localized radioactive sources in an exposure system where the ions are to be generated above the animal zone. To illustrate this problem Figure 1 shows the experimental setup for measuring the interaction between two plumes. The setup is similar to that described above except that two polonium sources (each consisting of a linear array of five patches) are mounted on the underside of the top electrode with their long axes separated by 6 cm. A 20-kV/m electric field was present in the 40-cm region above the ground plane and a 3-kV/m field was generated in the space above the screen electrode. The results of the scan are shown in Figure 2. The ion plumes generated by each radioactive source remained distinct. A more uniform space charge distribution near the ground plane than is shown in Figure 2 would require either more intense

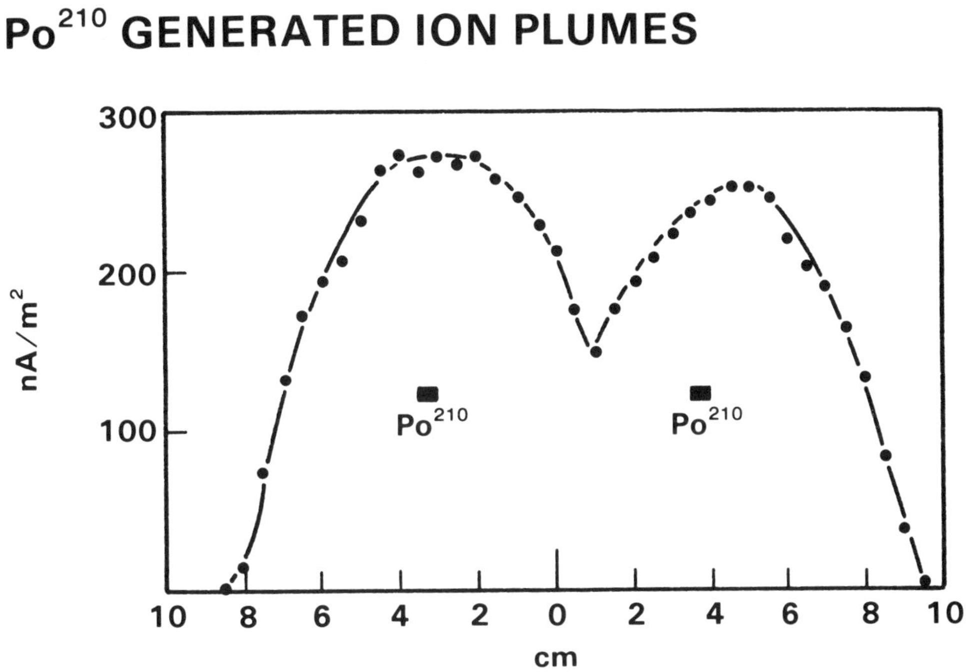

FIGURE 2. Scan of ground-level ion current densities generated by Polonium-210 sources. Squares labeled Po^{210} indicate the relative position and size of the radioactive sources. Abscissa is distance of probe along ground plane slot.

radioactive sources, sources spaced much closer together, or reduced electric field strengths. Each exposure system designer must evaluate whether his exposure level requirements and restrictions on radioactive sources in the laboratory will allow using this ion generation technique.

Corona discharge is also used in various laboratories to generate air ions. This technique has some disadvantages compared with radioactive decay. Corona sources generate noise that has a broad frequency spectrum extending to very high frequencies (>50 kHz)[10] while radioactive sources do not. The presence of noise in the exposure system is undesirable because the sham-exposed systems would either lack this noise or be required to generate a similar noise without introducing ions into the animal enclosure. Corona sources can also generate ozone, a potentially harmful agent. In tests at the Battelle exposure system where 68 μA of corona was being generated, ozone levels measured in an animal cage undergoing 10 air changes per hour were in the range of 1 to 20 ppb measured over a 24 hr period. This was the same range as measured in the exposure room with no corona present and is actually below the recommended time-weighted average-threshold-limit value of 100 ppb proposed by the American Conference of Governmental Industrial Hygienists. Finally, the electrical system needed to generate and control the corona-produced space charge is more complex and at a higher potential than for radioactive sources. Against these arguments is the fact that the use of radioactive substances on the relatively large scale needed for biological testing introduces additional safety questions and is more expensive, especially if the exposure area is extensive.

Both of the exposure systems discussed in detail in this chapter use corona sources. The Rockefeller system has a much smaller exposure zone, so it uses a single stainless steel needle electrode imbedded in phenolic plastic to generate the ions. The Battelle system uses 0.003″ tungsten wires; 14 strands are strung 6 cm apart on a 1 m^2 frame of copper tubing.

In both systems, additional metal screen electrodes at potentials between the corona electrode and ground are used to regulate and distribute the ions over the exposure zone and in addition, generate the electric field which directs the path of the ions.

C. Housing System Design

Any satisfactory design for animal caging in an exposure system should insure that the housing is itself not stressful to the animals and that it provides for animal care and environmental controls. These requirements are not difficult. However, the additional requirement that within the cage, or at least in the volume in which the animal spends most of its time, the level of exposure should be stable, uniform, and measurable makes the design process extremely difficult. This section discusses the major problems which must be addressed when designing a system.

Designers of air ion exposure systems might be inclined to adopt similar designs as those used in 60-Hz exposure systems. These cages are invariably made from dielectric materials. The condition that must be satisfied so that a cage does not perturb the exposure electric field (and ion density since it can be assumed that the air ions, in the absence of large air movement, follow the paths of electric field lines) is the complex conductivity of the cage be comparable to that of air; i.e., $j\omega\epsilon_o \approx \sigma + j\omega\epsilon$ where $j = \sqrt{-1}$, ω is angular frequency, ϵ_o is the permittivity of air, σ is the conductivity of the material from which the cage is made, and ϵ is the permittivity of the cage material. At 60-Hz, this condition is approximately satisfied by a number of plastic materials.

Cages for exposure to static electric fields cause field perturbations since the complex conductivity of air without air ions is zero at zero frequency but the cage materials have nonzero conductivities. The situation is even more involved when space charge is present, since the air is slightly conductive with a conductivity of $\mu\rho$, where μ is ionic mobility and ρ is space charge density. The only passive way to avoid cage perturbations of the electrical environment is to select the cage conductivity to roughly match that of air.

Dielectric materials are classified by their bulk conductivities and dielectric constants and by their surface conductivities. The first two quantities are stable, but the latter is highly variable depending on the relative humidity and surface contamination. Since animals soil cages with body oils, urine and feces, variable surface conductivity is a problem which becomes more severe as the time spent in the cage increases. The rate at which this problem develops will depend on the cage design and the species and age of the animals.

These problems were illustrated by a number of tests performed at Battelle. The arrangement for the first test is shown in Figure 3A. A hollow plexiglas hemisphere (radius 7.6 cm) was placed on the grounded electrode of a parallel-plate system which had a plate separation of 10 cm and a 2-kV potential difference. This generated a 20-kV/m field (with the hemisphere removed) which was measured with a field meter placed in the ground plane. For the test, the hemisphere was centered over the field meter and conductive paste coated the lip of the hemisphere to insure continuous electrical contact with ground. The results of the test are shown in Figure 4. The figure shows the field measured beneath the hemisphere under two conditions, carefully cleansed using ethanol, and soiled, a condition caused by rubbing the hands over both surfaces of the hemisphere. The first half of each plot shows the field after the voltage is applied. For both cases, the field immediately decays, but as the figure shows, the rate of decay for the soiled case is much greater. Presumably, the contamination has increased the surface conductivity of the soiled surface, allowing charge to migrate more rapidly under the influence of the applied electric field. The figure shows that most decay occurred during the first 7 min; however the decay continued at a slower rate than shown throughout a 4-hr period. The field at that time was 1 kv/m under the soiled surface.

When the voltage generating the 20-kV/m field was removed at the end of the 4-hr period,

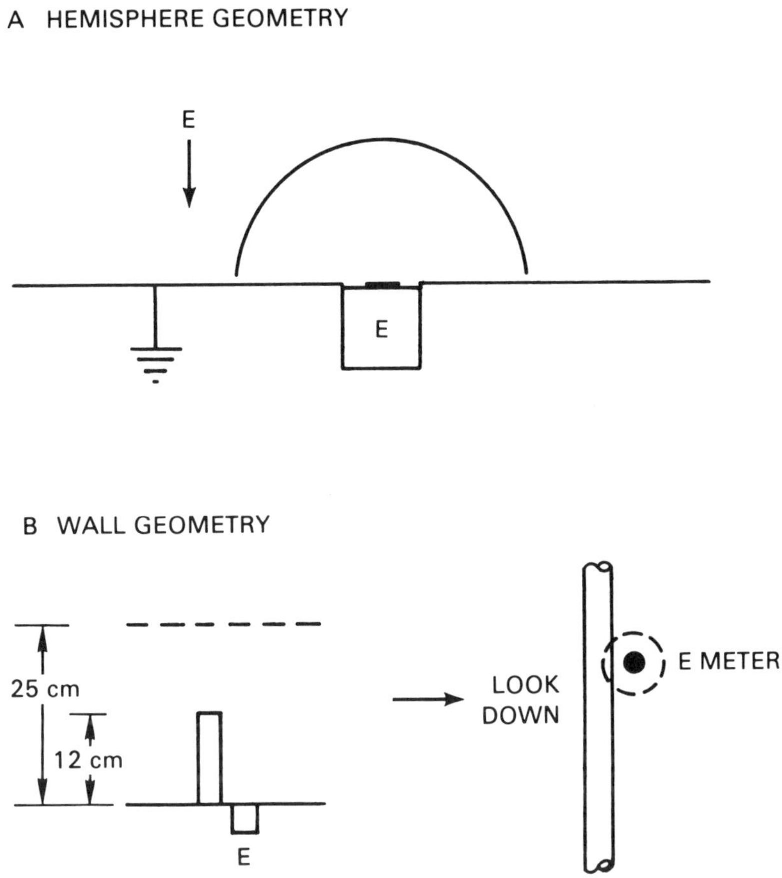

FIGURE 3. A. Arrangement for measuring the stability of the electric field beneath a plastic hemisphere. Electric field probe (E) is located in ground plane beneath the hemisphere. B. Arrangement for studying electric field and current density near acrylic plastic walls. Field probe (E) is in ground plane to the right of wall.

the field beneath both soiled and cleaned hemispheres reversed polarity and showed almost an identical rate of decay as seen during the on portion of the test. The polarity is reversed because the force generated by the field moves charge upward on the plastic hemisphere. This charge is opposite in polarity to that of the energized plate and, within a few seconds, neutralizes most of the field beneath the hemisphere (Figure 4). Similar results have been observed by Shigemitsu et al.[11] for a different experimental configuration.

Of course a hemisphere is not a typical animal enclosure. To illustrate the problem of using insulating material in a space-charge environment, the electric field at ground level near a 12.5 cm acrylic plastic (plexiglas) wall was measured 1 cm from the base of the wall, as shown in Figure 3B. The electrical environment was generated by a set of parallel plate electrodes similar to that described earlier in the study of ion generation technique. The electric field-generating electrode was energized to 7.5 kV and was separated from the ground plane by 25 cm. Electric field measurements were made over a span of 40 hr. During the first 20 hr, only the 30-kV/m electric field was present. At the end of the period, a 100-nA/m^2 current density was introduced into the test region.

Both cleaned and soiled walls were tested and the same basic pattern (Figure 5) was observed consistently. While the field alone is on, the measured field near the wall decays. Addition of space charge halts the decay, causing the field to rise and plateau at a level

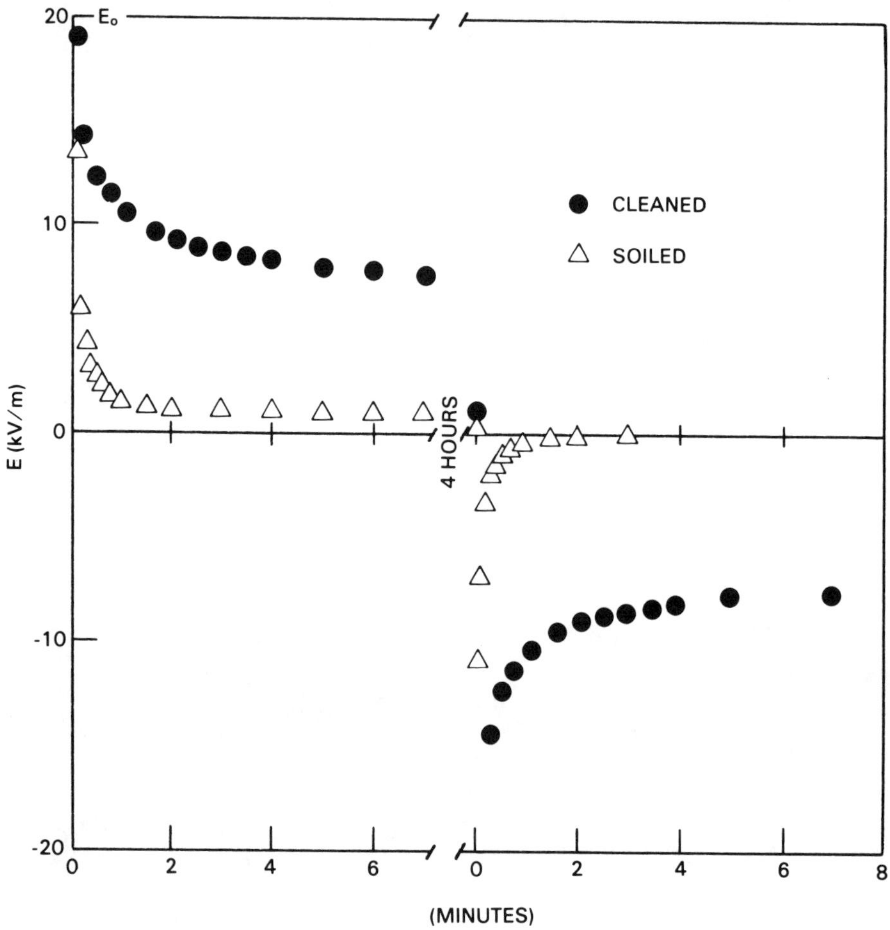

FIGURE 4. Decay of an electric field (no ions being generated) beneath an acrylic plastic hemisphere. During the time shown in the left side of each plot, a 20-kV/m field was being generated. The right side of the plot shows the residual field on the plastic immediately after the electrostatic field was turned off. Note there is a 4 hr gap between the two measurement periods shown.

generally above the unperturbed field strength. As expected, both the decay and rise to a new level were slower when the wall was cleaned than when it was not. An additional problem that is present when using an insulating wall is the effect of static charge even when no space charge is present. This effect was observed in the test shown in Figure 5. In cleaning the wall prior to beginning the test, a surface static charge was placed on the plastic. The charge generates its own field, which in this case, was measured to be about 15-20 kV/m. When the system was energized, the two fields from the wall and from the electrode add so that the total measured field was almost 50 kV/m. This additional charge decayed at a rate which varied with the cleanliness of the wall.

It is clear from these results that the traditional use of a cage made from insulating material, when placed in an electric field either with or without space charge, results in an unstable field whose magnitude and behavior are unpredictable. Even the presence of small amounts of such materials can affect regions nearby. This does not necessarily exclude all such materials but only points out the need for careful testing of the cage design. In fact, in the following section, two recently designed exposure systems will be described which use insulating materials in the construction of their exposure chambers.

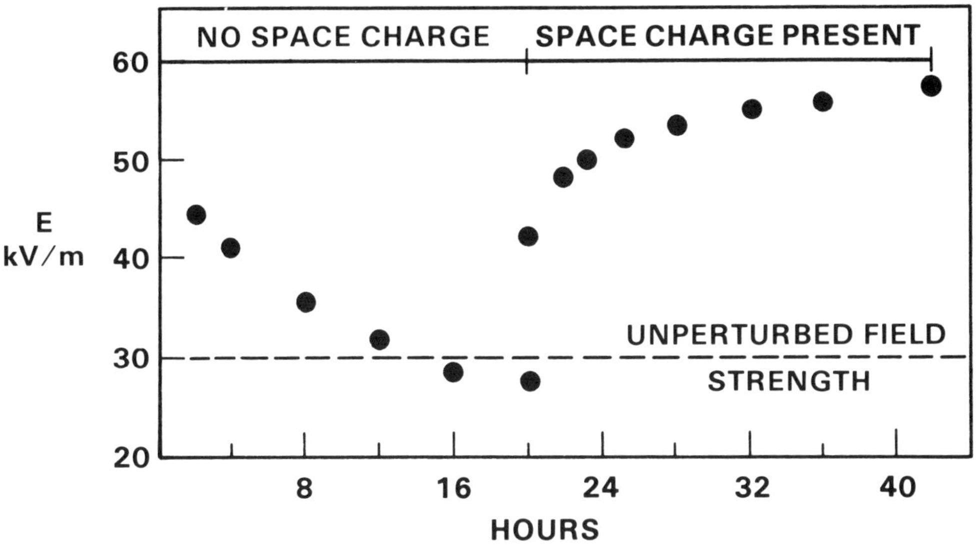

FIGURE 5. Electric-field measurements 1 cm away from cleaned plexiglas wall. During the first 20 hr, only a 30-kV/m field was generated. During the final 22 hr, in addition to the field, a 100-nA/m^2 current was present. The unperturbed level is that with the wall removed.

III. RECENTLY DESIGNED SMALL ANIMAL EXPOSURE SYSTEMS

A. The Rockefeller Exposure System

A carefully designed and characterized exposure system was developed at Rockefeller University by Charry and co-workers. It was originally described in 1983 and has since undergone a number of major modifications.[12] The Rockefeller chamber places great emphasis on delivery of clean, well-controlled air to the exposure zone. The air delivery system is illustrated in Figure 6. The delivered air is first humidified, and then filtered and cleaned by a course filter and then an activated charcoal filter. Finally a HEPA filter and electronic air cleaner insure particulate-free air. Temperature and relative humidity is held at 23° ± 1°C and 50% ± 5%. A blower is used to force air through the individual exposure chambers. Since the air-delivery system's pressure is positive with respect to the room, room air is kept out of the chambers. This design is an improvement over designs which only attempt to keep the room air clean since that air can never be as well controlled as in the confined area of the Rockefeller system. The delivery of clean air to the animals must accompany removal of fouled air and particulates which the animals generate by air flow through the cage. Air exchange in this system occurs at the rate of 4.4 changes/minute. Isolation of the exposure system environment from the laboratory in the Rockefeller system is insured by an outer chamber which encloses the exposure chamber. This outer section is sound insulated and electrically shielded. It is also mounted on cushions to absorb vibrations. In summary, this system illustrates the concerns which must be addressed in protecting the integrity of the system from environmental contamination.

The exposure system itself is designed to expose a single rat to air ions in a relatively small electric field. Exposures are generally less than 24 hr in duration (longest exposures have been 66 hr.) A number of individual exposure units can be connected to the air purification system. A single chamber is constructed of cylinders of stainless steel and Plexiglas, a combination of materials which at first glance appear to be the most unlikely of choices for an exposure cage but are empirically configured in such a way that the system produces an exceptionally uniform electrical environment. Figure 7 shows the block design

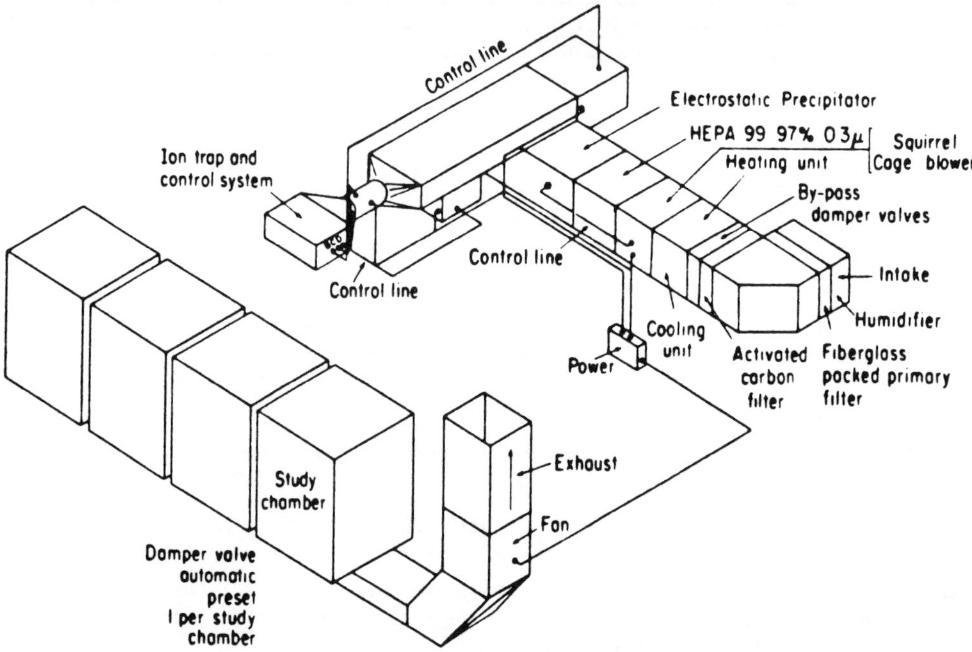

FIGURE 6. Air delivery and filtration system for Rockefeller system used to prepare air for delivery to the ion chamber.

of a chamber, which consists of five sections. These cylinders are stacked on top of each other in such a way that their interactions with the ions and with each other act to move the ions downward, with the help of the nearly 10 cm/sec air velocity. The first section, made of metal, couples the air delivery system to the chamber and evenly diffuses the air. Next, two cylinders of plexiglas contain the single ion-generating source and below that, a metal screen which controls the flux of ions by means of a voltage whose magnitude is adjustable between the corona electrode voltage and ground. The fourth section is the grounded stainless steel animal containment section (24.1 cm walls) with a stainless steel mesh floor. The lowest section contains the waste collection pan and air exhaust port. This section has no influence on the electrical environment but is essential for the maintenance of correct atmospheric conditions.

Looking at this design further, one can make some guesses as to the reason for its excellent performance but a theoretical analysis has never been done. Referring to the discussion on cage materials, we showed that ions collect on dielectric materials and generate a field which forces ions away from the walls. This undoubtedly occurs in sections 2 and 3 with a certain period being required before an equilibrium condition is established on the walls. Counteracting this force on the ions by the walls are three additional forces. The first is air movement which generates a downward force on the ions comparable to an electric field of approximately 600-700 V/m. Next is the electric field between the corona point and the electric field dispersion grid. This field has both a downward component and a radial component which tends to counteract the force generated by the static charge attached to the plastic walls. Finally, the grounded walls of section 4 form an equipotential surface which attracts the ions to the walls. The fact that these walls are grounded metal also means that animal contamination of the walls has no effect on the electrical performance of the exposure system. The system's developers have shown that the combination of these factors allows the ions to arrive at the ground plane uniformly distributed.

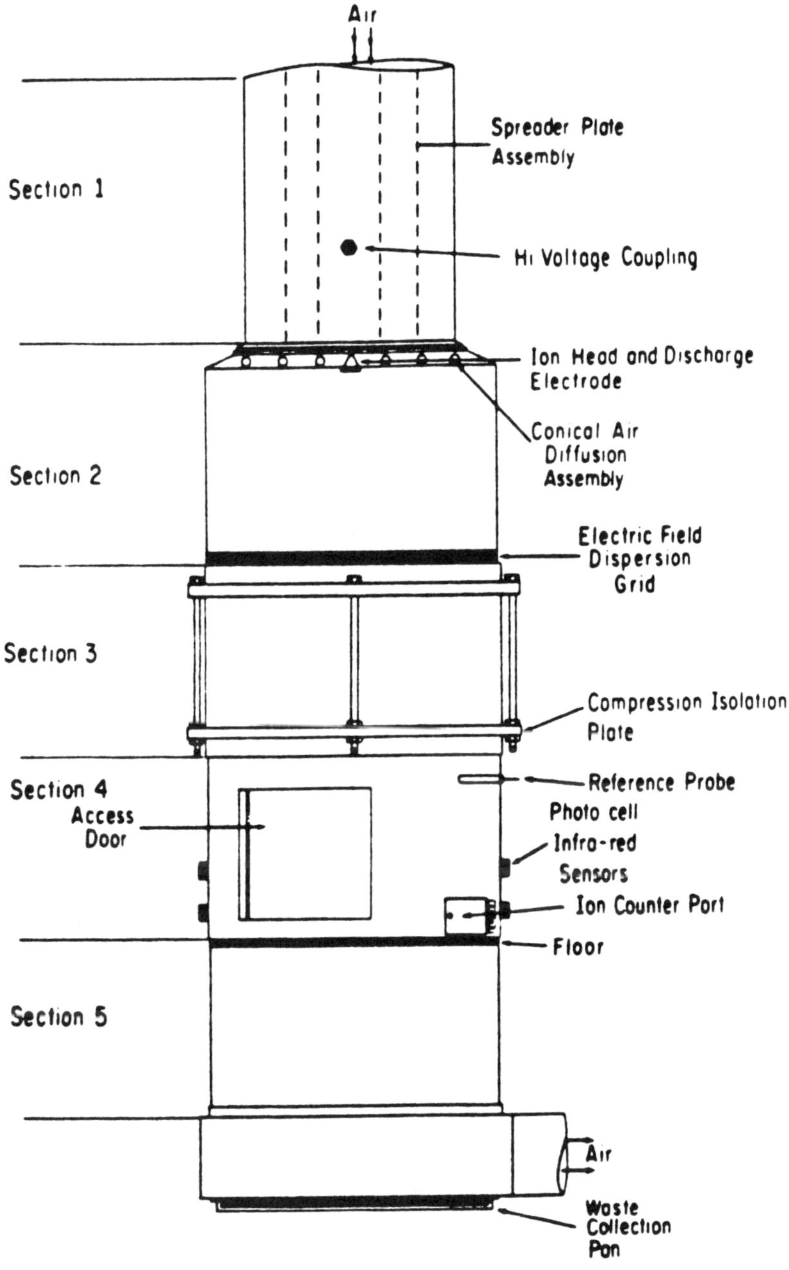

FIGURE 7. Components of the Rockefeller ion chamber.

B. The Battelle Small Animal Exposure System

A second air ion exposure system was developed by the author at Battelle Northwest Laboratories. This system had the following requirements which led to a very different design from the Rockefeller system. The environment which the Battelle system was to simulate was that found in the right-of-way beneath high voltage dc transmission lines. This environment is characterized by a static electric field that can be 30kV/m or larger, although generally it is considerably smaller. Air ion concentrations of either polarity are also variable, depending on the weather conditions and transmission line design, but levels of 10^5 ions/

cm^3 are not unusual. Thus the Battelle system must generate these large electric fields while most other air ion exposure systems generate an electric field only as result of the ion generating mechanism and the air ions themselves. This difference does not detract from its usefulness as an example of an exposure system because it is also capable of generating very low electric fields and high ion densities.

The other major requirement of the Battelle system was that the experiments require the simultaneous exposure of a large number of laboratory animals, usually rats, for durations of at least 30 days. To minimize costs and space utilization, individual exposure chambers were deemed to be impractical. Furthermore, environmental controls are room-wide (although the entire air supply to the room is HEPA filtered) instead of the carefully controlled environment so elegantly provided by the Rockefeller system.

The method of generating and controlling space charge is similar to that discussed in the section on ion generation techniques. Ions are produced by corona from fine wires strung on a 1 m^2 metal frame. Most of the ions are captured by one of three remaining electrodes: the cap electrode above the corona electrode and the regulating and field-generating electrode below it. Those ions which are captured form a current which flows to ground through a pair of resistors. The resulting voltage drop across these resistances energizes the three electrodes while the corona electrode is energized by the single high voltage dc power supply. The larger of the resistors (generally 200-300 MΩ) connects the field-generating electrode to ground and generates the voltage needed to produce the electric field in the exposure zone. The other resistor ($<$20 MΩ) is connected between the field-generating electrode and the other two electrodes (which are electrically connected). The small voltage drop across it is used to regulate the space charge concentration in the exposure zone. To obtain a specific combination of electric field and space charge density, both the magnitude of the corona current and the values of the resistances are adjusted empirically until the correct levels are obtained.

Cage design is the most critical part of the system. We have adopted an idea suggested by Frazier and Preache.[10] The cage walls are made from tempered glass which has a bulk resistivity approximately four orders of magnitude smaller than found in most plastics. The glass walls of the cages span the distance between the field-generating electrode and ground and are in physical and electrical contact with both. A small current flows through the glass and on its surface, producing a uniformly decreasing potential gradient on the walls which matches the space potential within the cage. Theoretically, the space charge will be neither attracted nor repelled from the wall, but will move vertically to ground.

Using the walls to actively stabilize the nearby space potential can only be expected to work if the bulk and surface conductivities of the glass are uniform and constant with time. For long-term animal studies, contamination of the walls, especially near the ground plane, can be expected to occur. This effectively shorts out the potential gradient on the wall and causes a decrease in the electric field in the region closest to the wall. To alleviate this problem, a number of 3 mm strips of copper tape with conducting adhesive were fixed to the outer surface of the cage, surrounding it at equally spaced intervals. Each of the strips is energized to a voltage equal to the theoretical space potential appropriate to its height above the ground plane. If a 30 kV/m field is to be generated in the cages, then the strip at 10 cm, for example, is at a voltage of 3 kV.

Figure 8 is a line drawing of one of the exposure modules. Eight glass cages are permanently fixed to a mobile cart, in a 4 $\times$ 2 configuration. Two mobile carts are placed under the set of electrodes. The cart and cages are sent through the cage washer as a unit since the cages are too heavy to move independently. The mobile carts serve a number of purposes besides a carrier for the cages. The waste pans are contained within the carts; the pressurized water delivery system delivers water to spring loaded nozzles located at ground level in the center of each cage. The room's air exhaust system is used to draw air down through the cage floor by way of a connection on the underside of the cart.

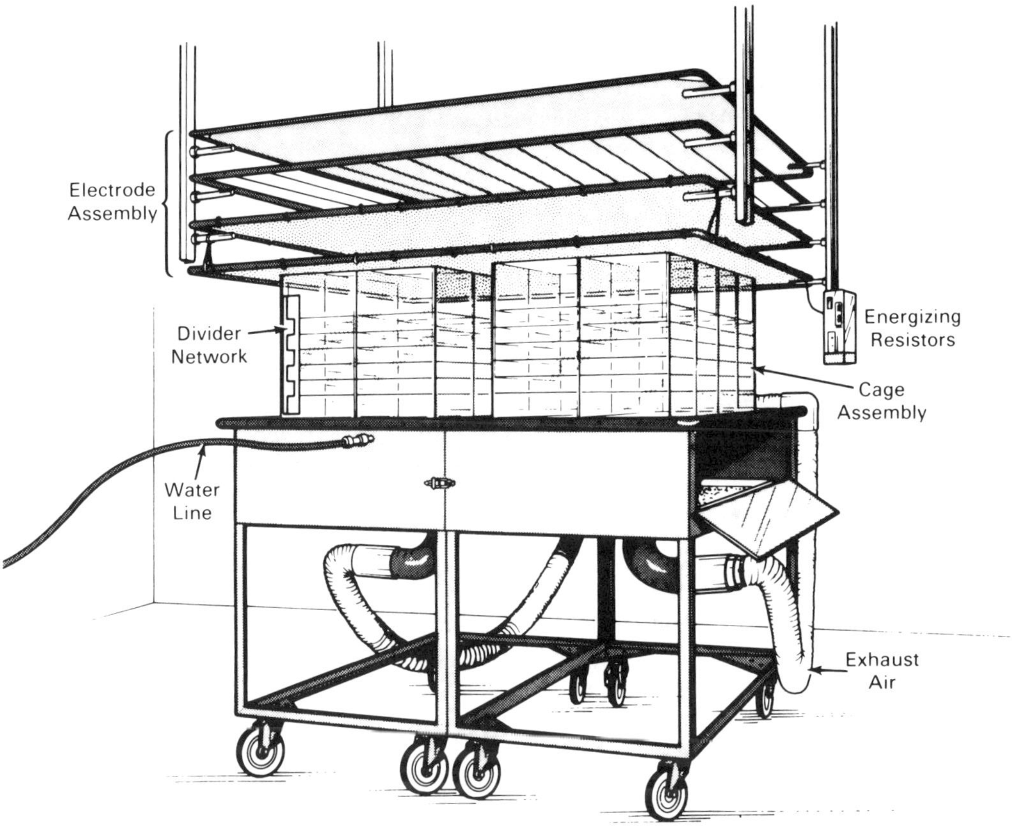

FIGURE 8. Sketch of a complete Battelle exposure module. Note the open access door to the waste catch pan, and the connections for water and exhaust. The field-generating electrode is suspended above the cages by straps at the front end to allow the electrode to be lifted in order to remove the cages.

IV. MONITORING EXPOSURE SYSTEM CONDITIONS

Careful documentation of an exposure system's performance is difficult to obtain but very critical. As we have discussed, it is hard to predict the paths of ions in any particular design. In addition, the electrical environment can be altered by such factors as relative humidity changes and contamination of the cage surfaces. As a result, besides periodic checks of the system's performance to insure that it is functioning correctly, it is extremely important to monitor the various electrical parameters throughout the duration of an experiment. Since in most designs, it is impossible to make these measurements while an animal is in the cage because it would interfere with or destroy the instrumentation, one should at the least periodically verify the exposure parameters in a cage at the end of the experiment (with the animal removed).

Instrumentation used in these monitoring programs have been described in the chapter by McKnight. However, the method of employing these instruments will often need to be adapted to the individual exposure system design. In general, field meters and Wilson plates (for measuring current density) must be mounted in the ground plane.

The Battelle system monitors the electric field and current density in the ground plane of an unoccupied cage throughout the duration of each exposure. A computerized data acquisition system samples and records these values and calculates an average exposure level. The time and duration of deviations beyond specified limits are noted. Ion density is not normally measured directly, but is calculated using the measured values for the field and current density and assuming an average ion mobility. Other monitored parameters are room temperature, relative humidity and the amount of corona current being generated by each set of electrodes. In addition to the computerized monitoring, scans have been taken of electric field and current density in a number of cages after 4 and 7 days of animal occupation. These measurements insure that the cage design can compensate for the contamination of the walls by the animal. Room particulate counts have been taken, and also ozone levels in the animal cages during corona generation. Finally, light levels and air flow through the cages have been balanced so that there is a minimum of variation between cages.

The Rockefeller system's electrical environment has been characterized using a calibration disc which is placed at ground level. The disc can be rotated 360° and the various measuring devices (ion counter, Wilson plate, and electric field probe) can be moved from the center of the chamber to the wall. Maps are made of each of the parameters at the ground plane. This has been done in both clean chambers and chambers in which an animal has resided for 18 hr. Ion density maps are obtained in two ways: (1) they are calculated, as they are in the Battelle system, using the measured values for current density and electric field; (2) direct measurements are made using a miniaturized aspirating-type ion counter. The results from the two methods agree reasonably well. Another unique feature of the Rockefeller system is the device used to monitor the ion current during an exposure. A metal rod is inserted horizontally into the cage 23 cm above the ground plane and out of reach of a rearing rat. The current collected by this rod is monitored continuously. From previous measurements, without a rat present, which calibrated the rod current to that measured with the calibration disc, the rod yields a useful indication of the level and stability of the ion current. Of course, fluctuations will be present as the animal moves about in the cage, but these can be averaged using a low pass filter.

V. INTERACTION OF AIR IONS WITH ANIMALS

In this section, we consider the question of "dose" of air ions received by the animal. This question cannot be answered at present because: (1) no one can precisely define "dose" since the location and mechanism of an effect has not been specified; (2) the accurate measurement of a location-specific dose is beyond the capabilities of presently available instrumentation. However, this does not preclude the experimenter from obtaining data to insure that the animal is being exposed at a level related in a predictable way to the unperturbed exposure levels generated in the exposure system. Such measurements are necessary to compare experiments in different systems where supposedly the "dose" is the same.

Obtaining dosimetric information about an exposure system can be simple or complex, depending both on the type of exposure system and the depth to which one wants to analyze the system. The most straightforward analysis can be performed in a system which generates a uniform electric field and space charge density and where the caging system does not alter the electrical environment. Such is the case for the Battelle system. For a system geometry which produces a nonuniform vertical field as would be the case where the ions and field are produced by a point source placed into corona, some of the procedures to be discussed in this section will apply only approximately. However, other described techniques for obtaining dosimetric information will still be applicable.

Consider a conducting object placed in a sinusoidal electric field with an angular frequency ω. If the body is grounded, then a current ("the short-circuit current") will pass from the body to ground. For uniform-field exposure, the situation may be pictured as follows: Associated with the field is a displacement current density, J, related to the electric field by the equation $J = \omega\epsilon_o E$. In the region above the body where the field is uniform, the displacement current density is uniform. However, at the surface of the body the field is

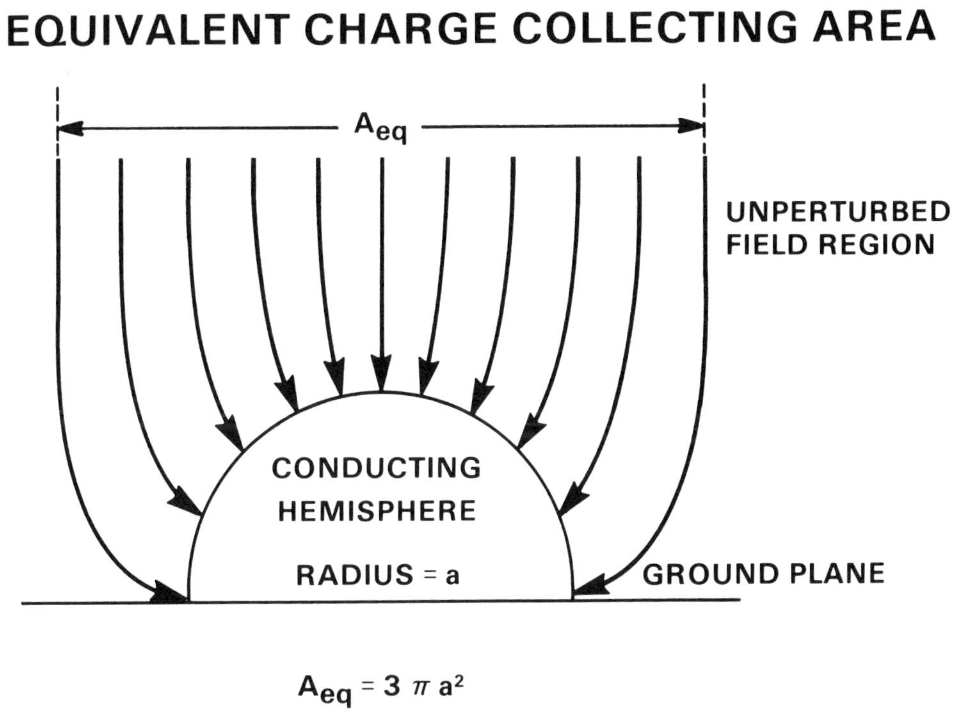

FIGURE 9. Representation of the equivalent charge-collecting area, A_{eq}, for a hemisphere with a radius **a** in a uniform electric field. Arrows represent field lines. For this simple geometry, it is calculated that $A_{eq} = 3\pi a^2$.

perturbed, is generally enhanced, and is perpendicular to the body.[13-16] This means that displacement current flows into the body. The total current entering the body is equal to the current leaving the body which, for a grounded body, is the short-circuit current. It is convenient to introduce the equivalent current collecting area, A_{eq}, defined by $A_{eq} = I_{sc}/\omega\epsilon_o E$, where I_{sc} is short circuit current and E is the unperturbed (uniform) electric-field strength. A physical interpretation of A_{eq} is shown in Figure 9. Note that field (and hence current-density) lines terminate on the exposed body if and only if they pass through A_{eq}. Thus, all the displacement current entering the body also passes through A_{eq}. Since A_{eq} is located where the field is uniform, the displacement current passing through it is easily calculated as $I = \omega\epsilon_o E\ A_{eq}$. Since $I = I_{sc}$ for a grounded body, this result is identical to the formula given above to define A_{eq}.

The electric-field distribution around a conducting body exposed to a dc electric field is identical to that which would be produced by an extremely low frequency sinusoidal electric field, provided that no space charge is present.[15] The situation becomes more complex when space charge is introduced because space charge also acts as a field source. For systems with space charge we assume that the space-charge distribution alters only the magnitude of the system electric fields, but not their directions.[17] This is only approximate but since, in the absence of winds, ions follow the electric-field lines, they will reach the surface of the hemisphere in Figure 9 if and only if they pass through A_{eq}. Thus, the ion current, I_c, collected by the body can also be related to the area A_{eq}:

$$I_c = J\ A_{eq} \tag{1}$$

where J is the *unperturbed* ion-current density. Ac and dc measurements of A_{eq} yield the same value as was shown at Battelle for both a conducting hemisphere and a hemiellipsoid at an electric field of 30 kV/m and ion densities from 10^4 to 10^6 ions/cm^3.[5] The measurement uses a device similar to a Wilson plate to measure current density, with either anesthetized animals, animal models, or geometric shapes. The body being studied is insulated from ground, usually with a thin piece of plastic (mylar) and connected by a short wire to the input of a high impedance current-measuring device. For anesthetized animals (or dead animals if muscle movements from breathing generate too much noise), an electrode is inserted into tissue (usually intraperitoneally) and connected to the input lead of the current meter.

For geometrical shapes, the equivalent current-collecting area can be calculated directly, but animals have a complex body shape which precules calculation of an ideal A_{eq} and it must be measured using a parallel plate electrode system to generate a 60-Hz electric field. The short circuit current (I_{sc}) is measured and the equivalent area, A_{eq}, is calculated as:

$$A_{eq} = I_{sc}/J_{ac} \tag{2}$$

where J_{ac} is the unperturbed displacement current density,

$$J_{ac} = \omega\epsilon_o V/d \tag{3}$$

where V is the voltage between the electrodes and d is the spacing between them.

Once the ac measurement is completed, the animal is transferred to the dc facility, where it is placed in an identical position on that grounded electrode with its body electrically connected to the dc current meter. Care should be taken to insure that the insulating material beneath the animal is cut to conform closely to its shape as protruding material could collect space charge and distort the electric field lines. The current collected by the animal's body is measured as are simultaneous measurements of the electric field and current density at a location away from the perturbing influence of the animal. The expected collected current from (1) and (2) is compared with the measured current.

These measurements give a value for the total number of ions per second which are impacting on the animal. This could be considered as a first approximation of the ion dose administered to the animal. The data also demonstrate whether the exposure system is delivering ions in a predictable way when the animal is present. In fact, if a conducting model of the experimental animal is positioned in a number of typical positions and postures within the cage, the measured current can be averaged. Any problems in ion delivery to the animal can be detected.

These data can answer an important dosimetric question raised by Frey:[7] Do the ions also accumulate on the fur of the animal? Such an accumulation could result in a reduced exposure of the subject because the field produced by ions on the fur would tend to deflect ions away from the animal. This effect is illustrated in Figure 10. Our measurements, which are still in progress, indicate that this can indeed be a problem, at least under our experimental conditions. Figure 11 plots the ratio of measured to calculated ion current against space charge density in the case where field strength with no ions present was 30 kV/m. The ion densities tested range from virtually zero to space-charge saturated levels.[18] The figure shows the results for three different animals: a large male rat, a much smaller female rat, and a guinea pig. For low space-charge levels, the measured collected ion current was close to that expected. However, as the ion concentration was increased, the ratio of measured to expected ion current progressively fell below 1 (in the case of rats, to a ultimate value less than 0.4). This effect can evidently be ascribed to charge accumulation on the fur of the live subjects. Clearly, an estimate of air-ion dosage based on ac dosimetric data could be a significant overestimate. Charge accumulation on fur is of course an effect that occurs only for fur-bearing animals and even then, the effect may depend on the species, sex, age and relative humidity. Thus, the above account should not be taken as an "ion effect" but as

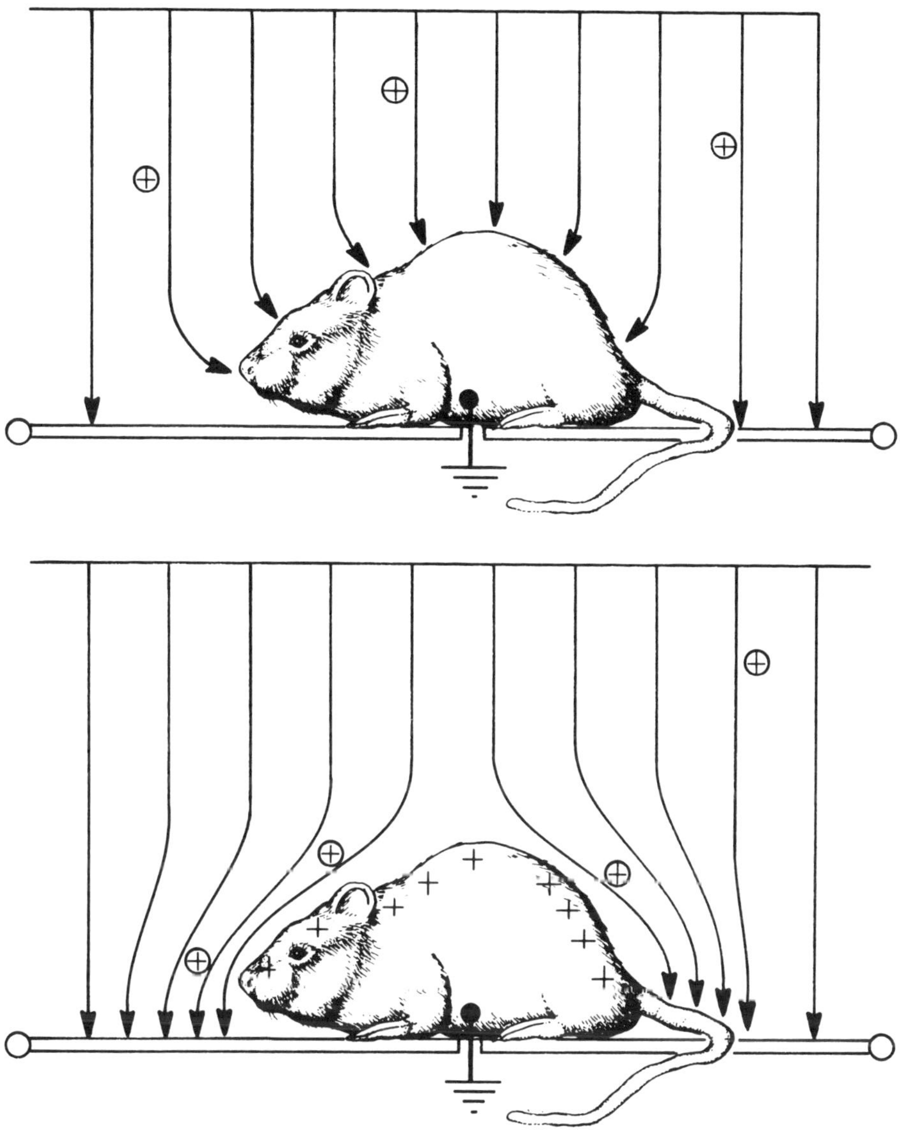

FIGURE 10. Conceptualizations of the possible behavior of ions (represented as +) in the presence of a rat resting on a ground plane in a uniform positive dc electric field. In the upper figure, the rat behaves like a conducting object in the electric field. Ions are collected by the grounded rat. The lower figure is an extreme example of what could occur if the fur of the animal were a good insulator. Charge is trapped on the fur, generating an electric field of its own which repels the remaining ions. Very little current would be measured.

a possible complication in determining dose. In comparing exposure data between species and in extrapolating data from an animal to human exposures, variables such as the fur effect must be considered.

This discussion of dosimetry methods is only a sampling of the kinds of measurements that can be made in characterizing the exposure. The exact tests that can be made depend on the design of the exposure system. Cage walls can have little or no effect on the shape of an ac field, but can significantly distort a dc field, thus the relevance of certain of the measurements mentioned above must be evaluated in the light of the specific cage design.

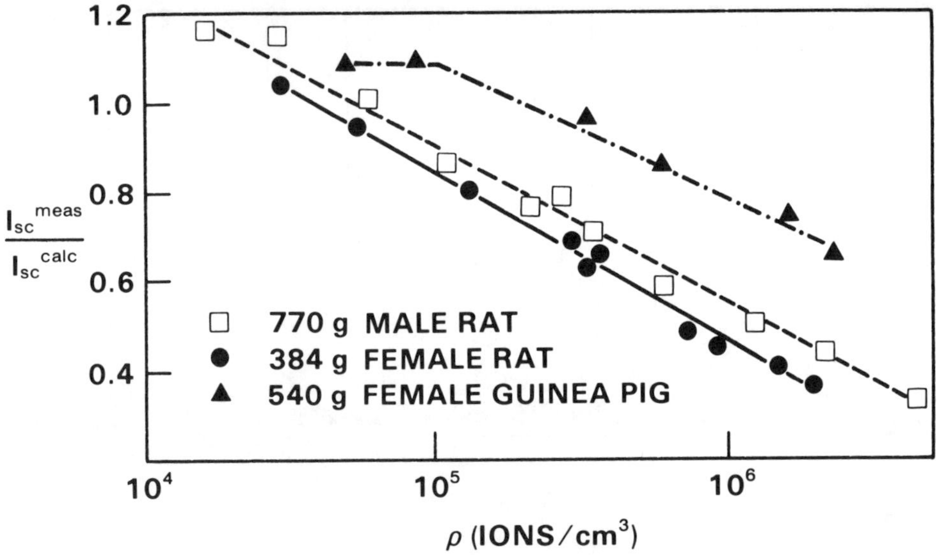

FIGURE 11. Results of short-circuit current measurements on three animals. Ordinate is the ratio of the measured collected body current to the current calculated using the equation $I_{sc}^{meas} = JA_{eq}$ where the value of A_{eq} is obtained from ac measurements and J is the measured current density with the animal removed. Ratios less than 1 indicate ion current is less than predicted if the animal were a good conductor.

When the electric field in the exposure system is small and due primarily to the space charge, then the collected current will be much smaller than discussed above. Care must be taken in that case to avoid measurement errors due to vibrations in the cable; a more sensitive current-measuring device will probably be required. However, the measurements made from anesthetized animals and from animal models can be valuable in detecting problems the system may have in delivering ions to the subject.

REFERENCES

1. **Andersen, I.,** Effects of natural and artificially generated air ions on living organisms, *Biometeorology,* 5, 229, 1972.
2. **Charry, J. M.,** Biological effects of small air ions, *Environ. Res.,* 34, 351, 1984.
3. **Krueger, A. P. and Reed, E. J.,** Biological impact of positive air ions, *Science,* 193, 1209, 1976.
4. **Chalmers, J. A.,** *Atmospheric Electricity,* Pergamon Press, Oxford, 1967.
5. **Weigel, R. J. and Kaune, W. T.,** Design, Construction and Testing of a DC Bioeffects Enclosure for Small Animals. Report EPRI EA-4189, Palo Alto, Calif., 1985.
6. **Bachman, C. H., McDonald, R. D., and Lorenz, P. J.,** Some physiological effects of measured air ions, *Int. J. Biometeorol.,* 9, 127, 1965.
7. **Frey, A. H.,** Modification of the conditioned emotional response by treatment with small, negative ions, *J. Comp. Physiol. Psychol.,* 63, 121, 1967.
8. **Gilbert, G. O.,** Effects of negative air ions upon emotionality and brain seratonin in isolated rats, *Int. J. Biometeorol.,* 17, 267, 1973.
9. **Krueger, A. P. and Reed, E. J.,** Effect of the air ion environment on influenza in the mouse, *Int. J. Biometeorol.,* 16, 209, 1972.
10. **Frazier, M. J. and Preache, M. M.,** Design, construction and testing of a dc bioeffects test enclosure for small animals, Report DOE/RA/10157-1, NTIS, Springfield, Ill., 1980.
11. **Shigemitsu, T., Tsuchida, Y., Nishiyama, F., Matsumoto, G., Nakamura, H., and Shimizu, K.,** Temporal variation of the static electric field inside an animal cage, *Bioelectromagnetics,* 2, 391, 1981.

12. **Charry, J. M., Bailey, W. H., Shapiro, M. H., and Weiss, J. M.,** Ion exposure chambers for small animals, *Bioelectromagnetics,* 7, 1, 1986.
13. **Deno, D. W.,** Currents induced in the human body by high voltage transmission line electric field-Measurement and calculation of distribution and dose, *IEEE Trans. Power Appar. Syst.,* PAS 96, 1517, 1977.
14. **Kaune, W. T. and Phillips, R. D.,** Comparison of the coupling of grounded humans, swine and rats to vertical, 60-Hz electric fields, *Bioelectromagnetics,* 1, 117, 1980.
15. **Kaune, W. T. and Gillis, M. F.,** General properties of the interaction between animals and ELF electric fields, *Bioelectromagnetics,* 2, 1, 1981.
16. **Kaune, W. T.,** Power frequency electric fields averaged over the body surfaces of grounded humans and animals, *Bioelectromagnetics,* 2, 403, 1981.
17. **Hill, H. L.,** *Transmission Line Reference Book HVdc to ±600 kV,* Bonneville Power Administration and Electric Power Research Institute, Palo Alto, Calif., 1976.
18. **Misakian, M.,** Generation and measurement of dc electric fields with space charge, *J. Appl. Phys.,* 52, 3135, 1981.

Chapter 6

BIOLOGICAL EFFECTS OF AIR IONS: A COMPREHENSIVE REVIEW OF LABORATORY AND CLINICAL DATA

Jonathan M. Charry

TABLE OF CONTENTS

I. INTRODUCTION

Reports in the literature indicate both wide-ranging and mixed effects of air ions on physiological and behavioral systems in animals and humans. No one particular biological system appears to be specifically receptive to air ion exposure, and, in fact, it appears that air ions, if they have an effect at all, would be best characterized as nonspecific environmental stimuli. However, not all researchers agree with this assessment, and, in fact, one investigator has recently proposed the presence of a specific receptor in the lung consisting of neuroendocrine cell bodies (see Chapter 8).

In any case, the air ion literature is difficult to interpret because of the uneven quality in experimental design and execution. This is important as any assessment of scientific data should take into account differences in quality. This chapter, then, has several purposes. The first is to update an earlier literature review.[1] The second is to provide a picture of the types of bioeffects that have been reported in the literature, and, in so doing, perhaps provide a direction for future research. The third is to indicate, in detail, the experimental methods and procedures that should be used to ensure scientific quality. The fourth and final purpose of this chapter is to provide a reasonably comprehensive cataloging of the biological effects reported and the methods used. Specifically, this review will attempt to identify those studies meeting minimally acceptable criteria for scientific quality, and, in so doing, provide an illustration of how these criteria may be applied.

The criteria necessary for valid scientific experiments to assess the potential for biological effects of air ions have been reviewed in detail elsewhere.[1] However, for convenience they will be reviewed here in summary form.

A. Criteria for Valid Scientific Experiments

Because of the contradictory nature of the data, the quality of the research carried out requires examination to identify those studies meeting minimally acceptable criteria for scientifically valid research. Those criteria are (1) control, measurement, and documentation of the physical environment; (2) experimental design and procedures and use of statistical methods for data analysis; and (3) replication in independent laboratories.

B. Control, Measurement, and Documentation of the Physical Environment

Characterization of the physical environment is important in both laboratory and epidemiological studies. In the case of epidemiological research, the best that can be expected is an adequate documentation of relevant variables such as air ion level, dc field strength, temperature and humidity. Laboratory research is quite a different matter, and its requirements are treated fully below.

The physical environment includes the test chambers or experimental rooms used for animal or human exposure to positive or negative small air ions. Control over the physical environment is essential because many biological and behavioral responses can be influenced by changes in temperature, humidity, and the presence of gaseous contaminants. A complete description of state-of-the-art test environments for air ion exposure can be found in Charry et al.[2] and Weigel and Kaune[3].

Aspects of the physical test environment requiring control and documentation are (1) electrical characteristics (electric field strength, current density or ion flux, ion concentration, ion polarity, and ion mobility); (2) by-products of electric field and ion generation such as ozone or oxides of nitrogen; and (3) the meteorological aspects of the ambient environment (temperature, humidity, pressure, and air velocity). Characterizing the various electrical parameters of the test environment allows a check on the accuracy of reported measurements since the relationship between current density, ion mobility, electric field, and ion concentration is well known and can be calculated.[4]

Other factors to be considered include measurement of gaseous contaminants given off as by-products of ion-generation. These are important because some of the biological and behavioral effects attributed to air ions may also be produced by exposure to ozone or oxides of nitrogen, e.g., headache, irritability, loss of attention, and fatigue.

Finally, consideration of the air handling system including an assessment of air filtration in particulate removal is important. If air filtration is inadequate, biological effects that may be observed cannot be attributed with confidence to the action of small air ions.

C. Experimental Design, Procedures, and Statistical Testing

Experimental design dictates the procedures to be used in setting up and running the experiment. Experimental design and procedures include consideration of (1) exposure period (length of time subjects were exposed to air ions); (2) levels of ions delivered to subjects; (3) grounding of the subjects; (4) randomization or counterbalancing, which concerns both the order in which dependent measures were taken and the order or sequence in which subjects were exposed to the different experimental conditions (e.g., ionized or ambient control); and (5) double- or single-blind observations. At the minimum, proper experimental design calls for the use of a control group or a repeated measures design, randomization or counterbalancing where appropriate, and single- or double-blind observations (to reduce bias on the part of experimenters). Without this, numerous artifacts may be present (carryover effects or confounding of dependent measures, threats to validity such as practice effects and history, and experimenter bias, to name only a few). The presence of these artifacts contests the validity of the experiment and makes it impossible to attribute, with certainty, to ions any effects observed.

Many reported studies do not employ statistical tests. Without such tests, it is difficult to assess the significance, validity, or reliability of reported data.

D. Replication

The external validity of experimental or clinical studies is dependent on whether reported effects have been replicated independently in experiments from other laboratories. Unless and until substantially similar data are produced again by others, a study cannot be relied upon as scientifically conclusive.

II. AIR IONS: SUMMARY OF ANIMAL AND HUMAN STUDIES

For clarity, the literature on air ion effects is divided into the following groups:

1. Human physiology and behavior — Tables 1 to 5
2. Animal physiology and behavior — Tables 6 to 10
3. Human biomedical responses — Tables 11 to 15
4. Animal biomedical responses — Tables 16 to 20

Two sections on physiology and behavior, one for animals and one for humans, are reviewed. Included are those works which studied changes in behavior observed as a result of exposure to air ions and the physiological responses related to these changes. The section on humans includes a discussion of findings regarding effects of air ions on electroencephalogram (EEG) activity, performance (e.g., reaction time and vigilance), and mood. The section on animals treats findings regarding respiration, EEG activity, performance and learning indicators (e.g., avoidance tasks and conditioned responses), and other responses including the production of serotonin, a neurotransmitter which regulates a number of behaviors such as aggression and sleep activity.

Of the two sections reviewed on biomedical responses, one concerns humans and the other, animals. Included are studies of the effects of ions on pain due to burns, respiratory disorders, urinary monoamines such as serotonin and norepinephrine, urinary steroids, and clinical symptoms such as headache and sore throat. The section on animals includes a review of a number of reports of research which primarily examined the effects of ions on respiratory infections and developmental growth.

A total of 89 articles on the subject of biological effects in animals and humans of air ions have been included in this review. These articles were selected as representative of the work available in the four areas mentioned earlier, and constitute a reasonably exhaustive review.

This review illustrates the inherent difficulties involved with conducting rigorous, properly controlled, scientific inquiry into the behavioral and biological effects of air ions. These problems tended to be most serious for the earlier research into ion effects (prior to 1950), and so those studies have generally not been reviewed here.[5] Among the studies reviewed here are those that have been most rigorously designed and implemented. The findings from the literature surveyed are summarized first, followed by a more detailed evaluation of those studies found to be minimally acceptable in scientific quality.

A. Human Physiology and Behavior

1. Reported Results: Aggregate Data

Thirty-three studies of the effects of air ions on human physiology and behavior are reivewed here and are shown in Tables 1 to 5. The results are summarized in Table 1. The experimental methods and procedures used in these studies are shown in Tables 2 to 5.

Research examining physiological functioning in humans (Table 1) has determined that neither positive nor negative ions have effects on such biological responses as pulse and respiration, glucose, steroids, and on some of the biogenic amines which are important in regulating cognition and emotion. The studies that reported effects from air ions generally observed that positive and negative ion exposure had opposite effects on behavioral and physiological responses. There are, however, numerous reports indicating no effects from air ions, and, in some cases, the information about effects for particular responses among the different studies is contradictory.

There are limitations on what can be said about the effects of small air ions on human physiology and behavior from the results of any one of these studies. These limitations are primarily attributable to study design and control problems which are discussed more fully below. There is no overall agreement among the findings concerning physiological or behavioral responses in humans exposed to air ions. Some data indicate that ion exposure influenced reaction time, vigilance, critical flicker-fusion (a perceptual discrimination task), EEG patterns, and galvanic skin response. On the other hand, there are also reports that positive and negative ion exposure had no effect on variables such as pulse, respiration, heart rate, muscle tension, nor on perceptual motor task such as reaction time, vigilance, and brightness discrimination (see Table 1). Some observations, however, can reasonably be made based on patterns that appear in the aggregate data from these studies.

Five studies suggest ion levels of approximately 10^3 ions/cm^3 are not likely to affect mood, performance, or physiological variables such as respiration, pulse and blood pressure. Three of five studies indicated in Tables 1 and 2 which used ion levels of this magnitude reported no effects from ion exposure.[6-8] The other two studies indicated improved performance from exposure to negative ions, while positive ions exerted no discernible influence.[9,10] The ion level used in these studies — 10^3 ions/cm^3 — is an ambient ion density commonly found in a variety of outdoor and indoor environmental settings.[11]

Six of the eight studies employing ion concentrations on the order of 10^4 ions/cm^3 showed some effects from exposure to ions of either polarity.[12-17] Positive ions were associated with increased reaction time and mood decline. Negative ions, by contrast, appeared to shorten

Table 1
SUMMARY OF EXPERIMENTAL RESULTS: HUMAN PHYSIOLOGY AND BEHAVIOR

Study	Physiological or behavioral response	Air ion effects[a] (+)	Air ion effects[a] (−)
Albrechtsen et al., 1978	Mental performance	0	0
	Pulse and respiration	0	0
Assael et al., 1974	EEG alpha		
	Frequency	–	↓
	Amplitude	–	↑
	Synchronization	–	↑
Barron and Dreher, 1964	Psychomotor performance		
	Brightness	–	0
	Reaction time	–	0
	Complex perceptual-motor	–	0
	Muscular steadiness	–	0
	Biochemical responses		
	Glucose, steroids & biogenic amines	–	0
Brandt, 1933	Metabolism	Bipolar Hi:	↑
		Lo:	↓
Buckalew and Rizzuto, 1982	Anxiety	–	0
	Psychological state perception	–	0↑
	Physiological state perception	–	0↑
Buckalew and Rizzuto, 1984	Psychomotor and cognitive performance		
	Digital symbol coding	–	0
	Grip magnitude	–	0
	Dexterity	–	0
	Tracking	–	0
	Reaction time	–	0
	Physiological responses		
	Body temperature	–	0
	Blood pressure	–	0
	Pulse rate	–	0
Charry and Hawkinshire, 1981	Galvanic skin response	↓↑	–
	Reaction time	0↑	–
	Mood	↓	–
Chiles et al., 1960	Complex mental task	0	0
	Vigilance	0	0
	Mood	0	0
DeVries and Klafs, 1965	Physical endurance	0	0
Forward and Alpern, 1980	Anxiety	–	↓
Giannini, 1979	Irritation syndrome	↑	–
	Anxiety	↑	–
Giannini, 1983	Irritation syndrome	↑	–
	Anxiety	↑	–
Hawkins and Barker, 1978	Performance		
	Motor tasks	0	↑
	Reaction time	0	↓
Hawkins, 1981	Thermal-comfort	–	0↑
	Mood	–	0↑
	Air quality rating	–	0↑
Halcomb and Kirk, 1965	Reaction time	↑	↓
Hedge and Eleftherakis, 1982	EEG alpha	–	0
	Reaction time	–	0
Inbar et al., 1982	Heart rate	–	↓
	Rectal temperature	–	↓
	Skin temperature	–	0
	Perceived exertion	–	↓

Table 1 (continued)
SUMMARY OF EXPERIMENTAL RESULTS: HUMAN PHYSIOLOGY AND BEHAVIOR

Study	Physiological or behavioral response	Air ion effects[a] (+)	(−)
	Sweat rate	–	0
	Minute ventilation	–	0
	Blood pressure	–	0
Kelley, 1963	Reaction time	0	0
	Movement time	0	0
	Accuracy — rotary pursuit	0	0
	Steadiness	0	0
	Grip strength		
Knoll et al., 1961	Reaction time	↑ ↓	↑ ↓
Koyranskiy and Ukvol'berg, 1961	Simple task completion	↑ ↓	–
	Complex task completion	↑ ↓	–
McDonald et al., 1967	Reaction time	0	0
	Vigilance	↑	↑
	Heart rate	0	0
	Respiration rate	↓	↓
	EMG	0	0
McGurk, 1959	Mood	0	0
Monaco and Acker, 1963	Mood	0	0
	Heart rate	0	0
	Respiration	0	0
	Skin temperature	0	0
	Muscle tension	0	0
	Skin resistance	0	0↑
	Blood pressure	0	0↓
	Pulse finger volume	0	0↑
Morton and Kershner, 1982	Memory	–	↑
	Attention	–	0↑
Rim, 1975	Performance — cognitive	↓	–
	Psychiatric symptomatology	↓ ↑	–
Ruocco, 1962	Reaction time	0↑	0↓
	Mood	0	0
	Placebo: well being	↓	↑
Sigel, 1979	Mood	0↑	0↑
	5-HIAA	0↑ ↓	0↑ ↓
Silverman and Kornblueh, 1957	EEG Alpha		
	Frequency	↓	↓
	Relaxation	↑	↑
	Sleepiness	↑	↑
	Dry mucosa	↑	↑
Slote, 1961	Reaction time	↑	↓
	Critical flicker fusion	↓	↑
	Finger tapping	↓	↑
Sovijarvi et al., 1979	Heart rate	↑	↓
	Perceived exertion	↑	↓
Tom et al., 1981	Performance		
	Reaction time	–	↓
	Rotary pursuit	–	0
	Mood	–	↑
Wofford, 1966	Reaction time	–	↓
Yaglou et al., 1933	Respiration rate	0	0
	Pulse rate	0	0
	Blood pressure	0	0
	Oral temperature	0	0

[a] – = not studied; 0 = no effects; ↑ = increase; ↓ = decrease.

Table 2
SUMMARY OF EXPERIMENTAL CHARACTERISTICS: HUMAN PHYSIOLOGY AND BEHAVIOR

Study		Air ion dosages (+)	(ions/cm^3) (−)	Exposure period (+)	Exposure period (−)
Albrechtsen et al., 1978		High 9.0×10^3	9.0×10^3	1 day	1 day
		Low 3.0×10^2	3.0×10^2	1 day	1 day
Assael et al., 1974		–	NR	–	45 min
Barron and Dreher, 1964		–	1.0×10^3	–	1.25 hr
Brandt, 1933		$82—5.9 \times 10^2$	Bipolar	1 year, once/week	
Buckalew and Rizzuto, 1982		–	NR	–	6 hr
Buckalew and Rizzuto, 1984		NR	NR	–	6 hr
Charry and Hawkinshire, 1981		$2.0—3.0 \times 10^4$	–	1.5 hr	–
Chiles et al., 1960		2.6×10^4	2.1×10^4	2 hr	2 hr
DeVries and Klafs, 1965		$2.3—4.5 \times 10^5$	$2.3—4.5 \times 10^5$	15 min	15 min
Forward and Alpern, 1980		–	NR	–	3 $^1/_2$ weeks, 8 hr/day
Giannini, 1979		NR	NR	NR	NR
Giannini, 1983		$1.1—3.8 \times 10^3$	–	Variable	–
Halcomb and Kirk, 1965		1.0×10^4	1.0×10^4	4.3 hr	4.3 hr
Hawkins, 1981	Site 1	0.5×10^2	2.5×10^3	–	
	Site 2	1.3×10^2	2.0×10^3	–	4—6 weeks
	Site 3	1.0×10^2	3.5×10^3	–	8 hr/day
Hawkins and Barker, 1978		7.0×10^3	7.0×10^3	2 hr	2 hr
Hedge and Eleftherakis, 1982		–	$0.9—1.5 \times 10^4$	–	45 min
Inbar et al., 1982		–	$1.4—1.9 \times 10^5$	180 min	180 min
Kelley, 1963		1.2×10^7	1.2×10^7	1 hr	1 hr
Knoll et al., 1961		$2.0 \times 10^3—1.0 \times 10^6$	$2.0 \times 10^3—1.0 \times 10^6$	1 hr	1 hr
Koyranskiy and Ukvol'berg, 1961		–	Series 1: 1.15×10^5	–	5 days 30 min/day
			Series 2: 1.0×10^6	–	5 days 30 min/day
McDonald et al., 1967		10^6	10^6	45 min	45 min
McGurk, 1959		8.0×10^3	8.0×10^3	2 hr	5 hr
Monaco and Acker, 1963		NR	NR	20 min	20 min

Table 2 (continued)
SUMMARY OF EXPERIMENTAL CHARACTERISTICS: HUMAN PHYSIOLOGY AND BEHAVIOR

Study	Air ion dosages (ions/cm^3) (+)	Air ion dosages (ions/cm^3) (−)	Exposure period (+)	Exposure period (−)
Morton and Kershner, 1984	–	$1.0—1.2 \times 10^4$	–	20 min
Rim, 1975	NR	–	NR	–
Ruocco, 1962	$3.4 \times 4.3 \times 10^4$	$3.4 \times 4.3 \times 10^4$	1+ hrs	1+ hrs
Sigel, 1979	1.0×10^5	1.0×10^5	2 hr	2 hr
Silverman and Kornblueh, 1957	NR	NR	30 min	30 min
Slote, 1961	2.0×10^4	2.0×10^4	15—25 min	15—25 min
Sovijarvi et al., 1979	9.0×10^4	9.0×10^4	NR	NR
Tom et al., 1981	–	1.6×10^4	–	15 min
Wofford, 1966	–	NR	20 min	20 min
Yaglou et al., 1933	$5.0 \times 10^3—1.5 \times 10^6$	$5.0 \times 10^3—1.5 \times 10^6$	0.5—3 hr	0.5—3 hr

Note: NR = not reported; – = not studied.

reaction time and enhance mood. The increases or decreases in reaction time (the periods of time elapsing between the presentation of a stimulus and a subject's response) that have been reported as statistically significant were on the order of milliseconds, a relatively small change in magnitude. The mood changes that occurred, for example, were increases or decreases in fatigue, concentration, and surgency (elation). These changes in mood were small in magnitude. Two studies indicated no effects from exposure to ions on mood or performance.[18,19]

The seven studies which used ion concentrations of 10^5 ions/cm^3 and above do not agree with one another and do not seem to follow from the generally consistent results which were obtained for lower concentrations. For example, one of these studies reported no effects from ions on physical endurance.[20] Two others reported effects for some variables, but not others.[21,22] A fourth and fifth study indicated effects of ion exposure for reaction time and performance that were variable and difficult to interpret.[23,24] A final study using ion concentrations of $10^5/cm^3$ reported no main effects of ions when analyzing data on the basis of exposure to either positive or negative ions alone. This investigation found some effects when data for unipolar ion exposures were pooled, although pooling data under these circumstances is questionable.[25]

The experimental results for very high ion exposures are an exception to the otherwise linear relationship between ion exposure and observed effects. Using ion levels of $10^3/cm^3$, few effects were reported for physiological responses or performance. At $10^4/cm^3$ effects were reported to occur with somewhat greater frequency, and, if effects of ions on biological responses were linear, higher ion levels would be expected to produce biological responses of greater magnitude or at least frequency. This is not the case, as reported effects at $10^5/cm^3$ are mixed. The cause of the apparent lack of linearity is unknown and any attempts to explain it would be highly speculative.

2 Critique of Experimental Design and Documentation

Tables 3 and 4 summarize elements of the physical environment requiring measurement and/or control in air ion research, and indicate the extent to which the physical environment was documented and controlled in each study. The first column in Table 3 indicates electric field strength — in all cases a dc electric field, expressed as volts per centimeter (V/cm). The second column lists current density (A/cm^2), and the third lists ion mobility (cm^2/Vs). Mobility (third column) is the rate at which a gaseous ion will travel in an electric field of 1 V/cm. Small air ions have mobilities ranging from 1.0 to 2.0 cm^2/Vs. The last column in Table 3 provides an indication of the air ion exposure system used in the particular study. In most cases description of the exposure system was limited.

In Table 4, the first four columns indicate temperature (degrees Celcius), relative humidity (%), barometric pressure (inches or mmHg), and ozone and oxides of nitrogen (ppm), respectively. In a number of cases, levels of ozone or NO_x were estimated either by qualitative comparisons with background concentrations or with calculations based on expected output and rate of air-exchange. In these cases, the label "Est" appears in column 4. The last column in Table 4 gives relevant details about the air-handling systems used to control the test environment. This column indicates the degree to which various aspects of the air-handling system were specified: type(s) of filter used (e.g., activated carbon, HEPA) and efficiency ratings for the particle removal capabilities of filters; rate of air flow through the exposure system; number of air changes per unit of time; and uniformity of airflow throughout the exposure system.

Only one of the 33 studies summarized in Table 3 reported electric field strength;[7] no study reported current density. Almost all studies reported the ion concentration used, but only six of these gave mobility data for the ions.

A further limitation, given the well-documented effects of heat and humidity on perform-

Table 3
DC ELECTRICAL ENVIRONMENTAL CHARACTERISTICS: HUMAN PHYSIOLOGY AND BEHAVIOR

Study	Field (V/cm)	Current density (A/cm²)	Mobility (cm²/V-s)	Ion exposure system
Albrechtsen et al., 1978	NR[a]	NR	0.8	Ions generated by alpha emitter suspended from ceiling 0.95 m above floor in laminar flow climatic chamber; ion measurement by cylindrical condensors
Assael et al., 1974	NR	NR	NR	Ions produced by corona discharge and measured with ion counter
Barron and Dreher, 1964	100	NR	NR	Ions produced by corona discharge and dispersed with fan; ions measured by "current between the plates of a concentric tube polarized capacitor"
Brandt, 1933	NR	NR	0.2	Normal fluctuations in outdoor and indoor ion concentration
Buckalew and Rizzuto, 1982	NR	NR	NR	Ions generated by corona discharge
Buckalew and Rizzuto, 1984	NR	NR	NR	Ions generated by corona discharge
Charry and Hawkinshire, 1981	NR	NR	1—2	Ions generated by corona discharge in climatic controlled chamber and measured wtih target ion collector
Chiles et al., 1960	NR	NR	>0.007	Ions generated by tritium foil (beta emitter) and corona discharge; ions measured wtih volumetric collector
DeVries and Klafs, 1965	NR	NR	NR	Ions produced by corona discharge
Forward and Alpern, 1980	NR	NR	NR	Ions generated by corona discharge and measured with volumetric ion counter
Giannini, 1979	NR	NR	NR	None
Giannini, 1983	NR	NR	NR	None
Halcomb and Kirk, 1965	NR	NR	NR	Ion generator and air conditioner in duct
Hawkins, 1981	NR	NR	1—2	Ions generated by corona discharge and dispersed by fan
Hawkins and Barker, 1978	NR	NR	NR	Ions produced by corona discharge and measured wtih ion counter
Hedge and Eleftherakis, 1982	NR	NR	NR	Ions generated by corona discharge and measured with atmospheric analyzer
Inbar et al., 1982	NR	NR	NR	Ions generated by corona discharge and measured with ionometer
Kelley, 1963	NR	NR	NR	Ions produced by corona discharge and dispersed with fan; ions measured wtih a current probe calibrated to a parallel plate ion counter

Knoll et al., 1961	NR	NR	NR	Ions generated by polonium (alpha emitter) ''separated by an electric field and accelerated by air flow''
Koyranskiy and Ukvol'berg, 1961	NR	NR	NR	NR
McDonald et al., 1967	NR	NR	NR	Ions produced by Kr 85 source transported by electric field and air flow at 42 l/min
McGurk, 1959	NR	NR	NR	Ions generated by corona discharge and dispersed by fan
Monaco and Acker, 1963	NR	NR	NR	Ions produced by corona or radioactive generator
Morton and Kershner, 1984	NR	NR	NR	Ions generated by corona discharge with target collector
Rim, 1975	NR	NR	NR	Fluctuations in ion count associated wtih weather conditions
Ruocco, 1962	NR	NR	NR	Ions produced by tritrium foil (beta emitter) and measured with parallel plate ion counter
Sigel, 1979	NR	NR	NR	Ions generated by corona discharge and measured with parallel plate collector
Silverman and Kornblueh, 1957	NR	NR	NR	NR
Slote, 1961	NR	NR	NR	Ions generated by tritium in air-conditioned room and measured with parallel plate collector
Sovijarvi et al., 1979	NR	NR	NR	Ions produced by corona discharge and measured wtih ion counter
Tom et al., 1981	NR	NR	NR	Ions generated by corona discharge and measured with ion counter
Wofford, 1966	NR	NR	NR	NR
Yaglou et al., 1933	NR	NR	1.3	NR

[a] NR = not reported.

Table 4
MICRO-CLIMATE ENVIRONMENTAL CHARACTERISTICS: HUMAN PHYSIOLOGY AND BEHAVIOR

Study	T (°C)	RH (%)	BP (in.)	O_3 (ppm)	NO_x (ppm)	Air delivery system
Albrechtsen et al., 1978	Var	expt 1: 35 expt 2: 50	NR[a]	NR	NR	60 air changes/hr, 1 complete air change/min
Assael et al., 1974	NR	NR	NR	NR	NR	NR
Barron and Dreher, 1964	NR	NR	NR	NR	NR	Air-conditioned and temperature-controlled room
Brandt, 1933	23	NR	NR	NR	NR	No air filtering, use of open window or air conditioner
Buckalew and Rizzuto, 1982	NR	NR	NR	NR	NR	Enclosed room with area of 37.2 m^2 with "self-contained cooling system"
Buckalew and Rizzuto, 1984	21	NR	NR	NR	NR	Windowless chamber with self-contained heating and cooling systems
Charry and Hawkinshire, 1981	22	44	NR	est.	NR	Large particle filter, electrostatic precipitator, temperature- and humidity-regulated vent system
Chiles et al., 1960	23	25—35	NR	NR	NR	Electrostatic precipitator; ventilation by air-circulating blower with "air change estimated to be at a rate of 100 cfm"; positive pressure maintained in test room
DeVries and Klafs, 1965	NR	NR	NR	NR	NR	NR
Forward and Alpern, 1980	NR	NR	NR	NR	NR	NR
Giannini, 1979	NR	NR	NR	NR	NR	None
Giannini, 1983	NR	NR	NR	NR	NR	None
Halcomb and Kirk, 1965	NR	NR	NR	NR	NR	Air conditioned and temperature controlled room
Hawkins, 1981	Site 1: 24.3 Site 2: 22.4 Site 3: 22.8	48.4 53.2 46.5	NR	NR	NR	Sealed air centrally handled in modern office building; no information provided on filtration, air exchange, etc.; O_3 levels not measured above ambient (levels not given)
Hawkins and Barker, 1978	NR	NR	NR	NR	NR	NR
Hedge and Eleftherakis, 1982	18—21	52—70%	NR	NR	NR	NR
Inbar et al., 1982	40 ± 1	25 ± 5	NR	NR	NR	Climatic chamber, otherwise NR
Kelley, 1963	NR	NR	NR	est.	NR	NR
Knoll et la., 1961	NR	NR	NR	NR	NR	Window open during testing, otherwise NR

Koyranskiy and Ukvol'berg, 1961	NR	NR	NR	NR	NR	NR
McDonald et al., 1967	NR	NR	NR	NR	NR	Compressed air supply, filtered, and flow regulated to 42 ℓ/min
McGurk, 1959	29	100	NR	NR	NR	NR
Monaco and Acker, 1963	NR	NR	NR	NR	NR	NR
Morton and Kershner, 1984	NR	NR	NR	NR	NR	NR
Rim, 1975	15	<25—30%	NR	NR	NR	Weather conditions
Ruocco, 1962	24—28	56—71	NR	NR	NR	NR
Sigel, 1979	25 ± 1.5	52 ± 4	29.84 ± 0.18	.002	NR	Air drawn through a ventilation duct and circulated wtihin test room by an electric fan
Silverman and Kornblueh, 1957	NR	NR	NR	NR	NR	NR
Slote, 1961	20—25	40—60	NR	NR	NR	Air-conditioned room (2000 cubic ft)
Sovijarvi et al., 1979	21—24	NR	29.09—30.16	≤0.5		Laboratory room with constant humidity and ventilation
Tom et al., 1981	40	45	NR	NR	NR	NR
Wofford, 1966	NR	NR	NR	NR	NR	NR
Yaglou et al., 1933	NR	NR	NR	NR	NR	Heated air, no filtering 1800 cfm

[a] NR = not reported

ance, is the lack of information on temperature, humidity and barometric pressure presented in these studies (see Table 4). Of the three-three studies, 18 gave no data on any of these factors, and 25 studies also failed to report on the presence of gaseous contaminants, specifically ozone or NO_x, which are by-products of corona discharge or radioactive emission. Moreover, in 14 of 33 cases, the air delivery system was not described, so that it is impossible to determine the extent to which various contaminants or pollutants may have been present (see Table 4). Of the remaining cases, only two provided enough detail to assure that clean air was being delivered to test subjects. For those studies in which the physical environment was poorly controlled or described, it is not clear how to interpret the results because it is uncertain whether observed effects were due to the action of small air ions and not some other factor.

Additional important aspects of experimental design are shown in Table 5. In most cases, the design was appropriate. Only 9 of 33 studies, however, reported that subjects were grounded, a factor that may influence whether subjects were actually exposed to the ions generated in the experiment. Moreover, ion level in the vicinity of the subjects can depend significantly on the electrostatic properties of clothing; in only a few cases were subjects' clothing described.

It is also noteworthy that very few of these studies have been replicated since the data of original publication, either independently or within the same laboratory. Several studies, however, have been partially replicated. Replication, as it is used here, refers to the use of the same or similar environmental conditions, experimental design, testing procedures, and dependent measures in subsequent experiments. A full replication is defined as a virtually exact duplication of an experiment by an independent laboratory. Partial replication is defined as similar findings reported by investigators who used similar, but not necessarily the same dependent measures, environmental exposure conditions, and experimental procedures. A "No" indication in the "Replication" column indicates either that no replication attempt occurred or that replication was attempted and a failure to confirm positive results occurred.

Six studies that examined the relationship between air ions and reaction time have been partially replicated (see Table 5). These studies indicated that positive ions were associated with slowed reaction time, while negative ions appeared to speed up reaction time. Because these studies suffered numerous experimental problems, it is not possible to ascertain how significant small air ions were as an agent responsible for altering reaction time. On the other hand, it should be pointed out that an almost equal number of studies (see Table 1) evaluated reaction time and found no effects due to ion exposure (for either polarity).

The far right-hand column of Table 5 contains an abbreviated enumeration of the various problems in each of the 21 studies. For example, "ion-sensitivity" of subjects was not considered in 17 of the studies. In four studies where sensitivity was considered, only one established it on the basis of reliable measures.[12] In another study, ion measurements were taken at an unspecified distance from subjects with no assurance that ion concentration was uniform throughout the large exposure area.[18] Previous work has shown that it is unlikely that such concentrations would be the same throughout an exposure chamber.[26] In a long-range study where ions were turned "on and off" no estimate of dosage was provided and there was no assurance, based on the experimental description, that "exposed" subjects actually were exposed or that all "exposed" subjects received the same dosage.[27] In two of the six studies reporting mobility, details related to its measurement were not reported, even though mobility figures were given; thus it is difficult to see how mobility was determined.

Overall, there are numerous limitations in the research regarding the effects of air ions on human physiology and behavior. In addition there are numerous examples of research supporting the null hypothesis as well as the "ion-effects" hypothesis. One way to evaluate a body of these kinds of data is to select specific responses to ions that have been studied and examine whether patterns in the data emerge. A review of Table 1 provides this infor-

Table 5
EXPERIMENTAL DESIGN, PROCEDURES AND STATISTICAL TESTS: HUMAN PHYSIOLOGY AND BEHAVIOR

Study	Design	Grounding	Statistics	Replication	Other experimental issues/problems
Albrechtsen et al., 1978	Repeated measures Counterbalancing Yoked pairs based on T selection	Yes	Friedman F test	No	No detoxification between treatments, no control for differences in sex or age, inadequate procedures for selecting sensitive subjects
Assael et al., 1974	Double-blind crossover	NR[a]	NR	No	Very slight EEG changes, though consistent in direction, were not subjected to statistical tests and may not be significant
Barron and Dreher, 1964	Repeated measures Double-blind	NR	F test	No	Query about (−) ion condition with (+) electrostatic field — unclear how space charge from negative ions resulted in a net (+) field; test subjects were healty, nonfatigued pilots; method of creating a void or ambient condition not explained
Brandt, 1933	Case stuides	No	NR	No	Small sample size (5 subjects); date of study makes ion measurements questionable
Buckalew and Rizzuto, 1982	Factorial Double-blind	NR	F test X^2	No	Ion measurement procedures unclear, possible confounding of expsoure due to simultaneous exposure of 12 subjects in one test area, basis of watching experimental with control groups probably inappropriate considering reference to Charry (1976), no assessment of smoker vs. nonsmoker differences given 6 hr exposure period
Buckalew and Rizzuto, 1984	Repeated measures Double-blind	NR	F test	No	No description of ion measurement procedures, use of young, healthy subjects (possible sample bias), no assessment of individual differences (ion-sensitivity) given reference to Charry and Hawkinshire, 1981

Table 5 (continued)
EXPERIMENTAL DESIGN, PROCEDURES AND STATISTICAL TESTS: HUMAN PHYSIOLOGY AND BEHAVIOR

Study	Design	Grounding	Statistics	Replication	Other experimental issues/problems
Charry and Hawkinshire, 1981	Repeated measures Single-blind Counterbalancing Ion sensitivity determined by ALS	Yes	Repeated measures F test	Partial	Mobility estimated and not measured
Chiles et al., 1960	Repeated measures Counterbalancing	NR	F test	No	Ion measurements not in subjects' breathing zone, likely presence of intermediate and large ions, confounding of exposure with three subjects exposed simultaneously, dielectric properties of clothing not specified
DeVries and Klafs, 1965	Repeated measures Counterbalanced	Yes	F test	No	Subjects exposed for 15 min and then tested in ambient conditions — exposure and test room at unspecified distances
Forward and Alpern, 1980	Repeated measures	No	t test	No	Field study taking place at work and therefore exposure period is variable
Giannini, 1979	Case study	No	NR	No	No direct evidence of ion exposure, confounding of effects, no experimental or case controls, small sample size (two subjects), questionable assumptions made about polarity of ion exposure
Giannini, 1983	Case study	No	NR	No	No description of ion measures and therefore no verification of exposure, no expiermental or case controls, small sample size (five subjects), confounding due to different environmental setting for each subject and use of anecdotal reports
Halcomb and Kirk, 1965	Independent two group	NR	F test	Partial	Lack of control group, ion measurement procedures questionable, question of

	Counterbalanced				whether measurements were taken in breathing zone of subject
Hawkins, 1981	Repeated measures	NR	F test X^2	No	Exposure period per week not reported (number of days or hours), insufficient explanation of data analysis procedures, effects of humidity and ions on headache confounded, no verification of uniform exposure for subjects within the test sites
Hawkins and Barker, 1978	Independent three group Random	Yes	F test	Partial	Ion measuring procedures not described, no counterbalancing of dependent measures, use of young male subjects limits generalizability, apparent error in reporting of ion levels
Hedge and Eleftherakis, 1982	Repeated measures Control-group	Yes	F test Mann-Whitney U	No	Questionable method for assessing ion-sensitivity by use of inappropriate and nonvalidated instrument for "weather-sensitivity"; no description of ion measurement procedures and no information regarding subject exposure
Inbar et al., 1982	Independent two group Single-blind	No	t test	No	Ion measurement procedures not described with no verification of actual exposure, possible confounding of exposure due to simultaneous testing of six subjects in one area. Use of very high temperatures (40°C) a potential moderator variable
Kelley, 1963	Repeated measures	Yes	F test	No	Limited generalizability due to sample population (young subjects)
Knoll et al., 1961	Repeated measures	No	X^2 t test	No	Levels of significance not reported, ion measurement procedures likely to produce artifact, design vulnerable to contamination and carryover effects, no counter-balancing mentioned
Koyranskiy and Ukvol'berg, 1961	Case Study	NR	NR	No	Evidence for placebo effects, highly inconclusive data, confounding by experimental demand characteristics, lack of fit between data and text

Table 5 (continued)
EXPERIMENTAL DESIGN, PROCEDURES AND STATISTICAL TESTS: HUMAN PHYSIOLOGY AND BEHAVIOR

Study	Design	Grounding	Statistics	Replication	Other experimental issues/problems
McDonald et al., 1967	Independent three group	Yes	F test	No	Physiological measures contain artifact due to inadequate and inappropriate habituation period
McGurk, 1959	Single group Repeated measures	NR	NR	No	Ion measurement procedures unclear, no rationale given for using different exposure periods for (+) and (−) ions
Monaco and Acker, 1963	Factorial Repeated measures Counterbalancing	NR	F test	No	Several implied variables (e.g., ion generator type, hospital status (patient/non-patient), drug/no drug), with no data showing as to whether these factors made any differences; no specification of age or sex, indication of ANOVA and F tests, but no actual numbers of P values given. Ion measuring procedures not described
Morton and Kershner, 1984	Factorial Double-blind	NR	F test	No	No description of ion measurement procedures, possible confounding of exposure, no verification of exposure, use of target ion collector does not discriminate small ions
Rim, 1975	Independent two group	No	NR	NR	Reporting of p levels but no identification of statistics used, rationale for tests selected not given, no information regarding exposure to ions, data confounded as Sharav conditions consist of changes in temperature, humidity, and ionization
Ruocco, 1962	Repeated measures	NR	F test	No	Small sample size limits generalizability, total ion exposure period unclear (minimum exposure was 1 hr adaptation prior to testing)
Silverman and Kornblueh, 1957	Repeated measures	NR	NR	No	Inconsistent procedures for introducing

					subjects to ions, Ion measuring procedures not described
Slote, 1961	Repeated measures	NR	F test	Partial	No indication of counterbalancing, very brief exposure period
Sigel, 1979	Independent two group Repeated measures	Yes	F test	No	Questionable basis for selecting ''weather-sensitive'' subjects, likely confounding of data due to lack of ''detoxification period'' between positive and negative ion exposures resulting in potential carry over effects, questionable procedure of polling positive and negative ion data in analyzing for effects, no effects found for a majority of mood items and meaningfulness of a.m. and p.m. differences in ion effects unclear
Sovijarvi et al., 1979	Repeated measures Double-blind Cross-over	Yes	Paired t-test Wilcoxon test	No	Limited description of ion measurement procedures, no control group, unclear whether subjects were exposed in groups or individually, limited generalizability due to use of young subjects in sample, questionable assumption about exposed skin area being related to ion exposure
Tom et al., 1981	Post-test only Double-blind	Yes	t test	Partial	Total exposure time not specified, scoring of reaction time task is unclear, extremely high ambient temperature (40°C or 104°F) is a potential confounding factor and may explain why four out of seven measures showed no effects, No assessment of ion sensitivity, considering reference to Charry (1976)
Wofford, 1966	Independent two group	NR	t test	Partial	Lack of counterbalancing with dependent measures raising question of possible order effects within each treatment condition

Table 5 (continued)
EXPERIMENTAL DESIGN, PROCEDURES AND STATISTICAL TESTS: HUMAN PHYSIOLOGY AND BEHAVIOR

Study	Design	Grounding	Statistics	Replication	Other experimental issues/problems
Yaglou et al., 1933	Repeated measures	No	NR	No	Contamination of physiological measures due to uncontrolled diet, variability of environmental conditions

[a] NR = not reported

mation. For example, eight studies found instances of effects of ions on reaction time.[9,12-15,17,23,28] By comparison, six studies found instances of no effects on reaction time.[7,9,19,22,29,30] Examination of the area of cognitive and psychomotor performance indicates that in most instances no effects of ions were found. Table 1 reveals 22 instances where no effects were reported and 14 where effects were reported. The numerical weight of the aggregate data here favors an interpretation of no ion influence on psychomotor (cognitive) performance. All studies examining pulse found either no effects or mixed effects.[6,29,31,32] The same is true for studies examining blood pressure.[21,29,31,32]

3. Results of Minimally Acceptable Studies

In the absence of established exposure limits or guidelines for air ions the scientific and technical literature becomes the only basis on which to evaluate and understand the effects of exposure. As pointed out previously, this body of literature is deficient in numerous respects. However, one possible use of this literature is to examine only those studies employing comparatively the best controls, design, and so forth. The criteria used for selecting these latter studies are

1. Reasonable control over and description of the physical environmental exposure conditions
2. Use of statistical tests where effects were reported
3. Lack of major flaws in experimental design and procedures

Studies meeting these criteria are likely to provide the most valid findings, and are reviewed in greater detail here. The nature and magnitude of the effects being reported are emphasized. Identification of the magnitude of the response is approached in two ways whenever possible; (1) absolute change or difference due to ion exposure and (2) percent change or difference. These data show overall that the effects of ion exposure are small in magnitude and within the range of ordinary biological variability.

Nine studies of ion effects on human physiology and behavior (see Table 1) merit further consideration based on their methodology and reporting of information. In the first,[6] no effects of ion exposure (either positive or negative at 9.0×10^3 ions/cm^3 for 1 day) were found on mental performance or physiological responses. Mental tasks consisted of tests of addition, use of visual cues and logical thinking. The physiological measures were pulse, respiration, and body temperature.

Similar outcomes were reported in the second study.[18] This research tested the effects of ions on measures of performance and mood indicating no effects from exposure to either positive or negative ions (2.6×10^4 ions/cm^3) for 2 hr. Measures of performance consisted of vigilance and a complex motor task. Mood indicators included anxiety, aggression, and depression.

In the third study,[12] exposure of human subjects to positive ions (2.0 to 3.0×10^4 ions/cm^3) for 1.5 hr resulted in effects on mood, performance, and physiological responses. Positive ions were found to affect some types of moods, for example: decreased sociability (gregariousness or friendliness), increased tension (feelings of tightness or nervousness) and increased fatigue (the sensation of feeling tired). The magnitude of these effects was small (on average there was an 18% change due to ion exposure). Mood changes found not to be affected were anxiety and agression. The performance measure used was simple visual reaction time, which tests the subject's ability to respond to the presentation of a visual stimulus. (The time elapsing between the presentation of a visual stimulus and the subject's correct response is referred to as reaction time.) No effects on performance or physiological activation were observed unless subjects were separated on the basis of ion-sensitivity (using a measure of autonomic nervous system reactivity called the autonomic lability score —

ALS). Subjects were grouped according to whether their ALSs were above or below the mean of the distribution.[1] Those below the mean, i.e., low ALS subjects, were found to be most consistently affected by exposure to ions. For low ALS subjects exposed to positive ions, reaction time was slowed down by 14 msec (0.014 sec), a change of 5% due to ion exposure. The change in physiological activation for low ALS subjects was 16%.

In the fourth study,[10] office workers were exposed to negative ion concentrations (an average of $2.7 \times 10^3/cm^3$ for 4 to 6 weeks, 8 hr/day). Hawkins found that negative ions generally were associated with a perception of increased environmental comfort (i.e., warmer and less stuffy) and enhanced self-comfort, but these effects did not occur for all work areas studied nor for all shifts of work. For these measures, ions increased perceived warmth by 12 to 13%, decreased stuffiness by 10%, and increased alertness by 4%.

The fifth study meeting the standard of minimally acceptable quality reported no effects of negative ions ($1.5 \times 10^4/cm^3$ for 45 min) on EEG alpha or on reaction time (positive ion effects were not researched).[19] In this study, subjects were assessed for ion-sensitivity using ''weather-sensitivity'' scores, a headache questionnaire, and a personality inventory. In addition, both simple and complex reaction time were used as measures of performance.

The sixth study exposed subjects to negative ions ($1.9 \times 10^5/cm^3$) for 3 hr and reported effects on heart rate, rectal temperature, perceived exertion and systolic blood pressure.[21] The magnitude of effects was a 9.3% change in heart rate, a 3% change in rectal temperature, and a 5.6% change in perceived exertion. No effects were found on skin temperature, sweat rate, minute ventilation, or blood pressure.

The seventh study meeting criteria for quality exposed subjects for 5 hr to 8×10^3 positive or negative ions. No effects were observed on any of several indicators of mood or performance.[8]

The eighth study meeting criteria of minimal acceptability in quality exposed subjects for 2 hr to 1×10^5 ions/cm^3 (either positive or negative).[25] Sigel reported no effects for five out of seven mood indicators (i.e., tension-anxiety, depression-dejection, anger-hostility, fatigue-inertia, confusion-bewilderment. For two such indicators — vigor and friendliness — significant differences among ion-exposed and control groups were reported. The data indicated that over time (during the course of this experiment) both vigor and friendliness declined significantly for controls while ion-exposed subjects showed no change. This was interpreted by Sigel as indicative of biologically ''stabilizing'' effects of air ions. Sigel also studied 5-HIAA in urine (a metabolite of serotonin thought to be involved in mediating ion effects) and reported mixed effects. The data for urinary 5-HIAA are internally inconsistent, with data showing instances of no effects as well as effects. One trend that did appear consistent indicated that both concentration and amount of 5-HIAA in urine for ion-exposed subjects increased over time, while control subjects showed decreased urinary 5-HIAA. However, it should be noted that this effect was entirely dependent upon experimental period. (Experimental periods were divided into morning and afternoon session.) These differences were observed in the afternoon, but not the morning experimental period. Interpretation of these data suggests weak effects highly dependent upon complex interactions with other variables.

The ninth study reported effects of ions on various psychomotor tests for subjects exposed either to positive or negative ions ($2.0 \times 10^4/cm^3$) for 25 min.[15] For all measures taken, the magnitude of the effects observed was small. For reaction time, Slote reported an 8.4% difference between subjects exposed to positive ions and controls and a 9.6% difference between negatively ionized subjects and controls. For critical flicker-fusion, differences were 12.9% (positive vs. control) and 4.2% (negative vs. control), respectively. For finger-tapping, a 5.7% difference between negatively exposed subjects compared to controls was observed, with no effects for positive ion exposure.

In summary, these studies meeting criteria for minimally acceptable quality indicate several patterns. Subjects exposed to ion concentrations of approximately 10^3 ions/cm^3 ranging from 25 min to 6 weeks (8 hr/day) showed no effects on mental performance, physiological parameters (such as pulse and respiration), and mood. Some effects on perceived comfort were observed, although these effects were lesser in magnitude than those produced by slight to moderate changes in temperature or humidity. At ion concentrations of 10^4 ions/cm^3 and exposure periods of 1.5 to 2 hr, psychomotor performance was influenced in several cases by positive and negative ions, and other cases was not. In all cases, effects observed due to ions were small in magnitude.

B. Animal Physiology and Behavior

The studies conducted to determine effects of air ion exposure on the physiology and behavior of animals are covered in Tables 6 to 10. The term animal is used here to refer to laboratory animals, such as albino rats, mice, guinea pigs, and rabbits. In all cases, laboratory animals are bred specifically for the purposes of experimentation. Animal experiments are often conducted to provide models for physiological or behavioral changes that might occur in humans. Extrapolation from animal data to humans, however, is always difficult and must be done with care.

A total of 28 studies of the effects of air ions on animal physiology and behavior were reviewed. These results are summarized in Table 6. The experimental characteristics and procedural elements of these studies are presented in Tables 7, 8, 9, and 10, respectively.

1. Reported Results: Aggregate Data

The data presented in Table 6 indicate that exposure of laboratory animals to positive air ions tended to decrease activity, decrease central nervous system arousal (as evidenced by EEG changes), inhibit learning and performance, increase blood serotonin levels, decrease brain serotonin levels, and increase sensitivity to pain. With the exception of brain 5-HT effects, negative ions appeared to exert effects which were generally opposite to these. Two recent and well-controlled studies examined three neurotransmitter systems,[33,34] including serotonin (5-HT), in several different brain regions and reported no effects of exposure to either positive or negative ions.

Table 7 presents the ion levels and exposure periods used for each of the studies of animal physiology and behavior. This table indicates that the most typical ion density for the animal studies is on the order of 10^5 to 10^6 ions/cm^3, higher than ion densities generally used for the studies of human physiology and behavior. The exposure periods used for the animal studies are variable, ranging from 30 min to 100 days, but this does not seem to affect the consistency of reported results. For example, three studies of the effects of negative air ions on levels of brain 5-HT (serotonin) used exposure periods ranging from 12 hr to several days. The results, however, are quite similar (see Tables 6 and 7), in that brain 5-HT levels are decreased in the presence of positively or negatively ionized air.

Several different aspects of animal physiology and behavior have been studied. These include conditioned emotional response (CER), learning and performance, serotonin levels, pain and analgesia, and physiological arousal and activity.

a. Conditioned Emotional Response

As indicated in Table 6, two studies examined the effects of ions on conditioned emotional response (CER) in rats.[35,36] CER is usually interpreted to be experimentally induced "fear" in animals. Both studies reported that exposure to negative air ions reduced CER, i.e., reduced fear in animals. These results are obtained despite the use of different exposure periods by the two studies (see Table 7). One of these studies did not report ion concentration,[35] so comparison of the studies on this basis is not possible.

Table 6
SUMMARY OF EXPERIMENTAL RESULTS: ANIMAL PHYSIOLOGY AND BEHAVIOR

Study	Physiological or behavioral response	Air ion effects[a] (+)	Air ion effects[a] (−)
Bachman et al., 1965	Heart rate	↑	↑
	Respiration	↑	↑
Bachman et al., 1966a	Activity	↑	↑ ↓
Bachman et al., 1966b	ECG	↑ ↓	↑ ↓
Bailey et al., 1983	Brain 5-HT	0	0
	NE	0	0
	DA	0	0
Bailey and Charry, 1984a	Brain 5-HT	0	0
	NE	0	0
	DA	0	0
Bailey and Charry, 1984b	Spontaneous activity	0	0
Bauer, 1955	expt 1: Maze — learning	0	0
	expt 2: Nest — building	0	0
Diamond et al., 1980	Cortical 5-HT	–	↓
	Cortical cylic AMP	–	↓
Dowdall and de Montigny, 1983	5-HT responsiveness	0	↑
Duffee and Koontz, 1965	Performance	0	↑
Falkenberg and Kirk, 1977	Avoidance learning	↓	↑
Frey, 1967	Conditioned emotional response	–	↓
Gilbert, 1973	Reactivity	–	↓
	Brain 5-HT	–	↓
Herrington and Smith, 1935	Weight gain	–	0
	Activity	–	↑
	Hemoglobin	–	0
Jordan and Sokoloff, 1959	Maze learning	–	0 ↑
Krueger et al., 1963	Blood 5-HT	↑	0
Krueger et al., 1966	Blood 5-HT	↑	–
Krueger et al., 1968	Blood 5-HT	↑	↓
Krueger and Kotaka, 1969	Brain 5-HT	↓	↓
Lambert and Olivereau, 1980	Avodiance learning	↓	0
	Weight changes	0	0
	Spontaneous activity	0	↑
Lambert et al., 1981	EEG hippocampal theta		
	Frequency	↓	↑
	Amplitude	↑	↓
	EEG cortex		
	Frontal amplitude	↑	–
	Occipital amplitude	↑	–
McDonald et al., 1965	Heart rate	↓	↓
	Respiration	0	0
Mizusawa et al., 1981	Avoidance learning	0	↑
	5HT	0	0
Nazzaro et al., 1967	Conditioned emotional response	↓	↓
Olivereau, 1970a	Pain-sensitive defense reactions	–	↓
Olivereau, 1970b	Pain-sensitive defense reactions	↑	–
Olivereau, 1979	Spontaneous activity	↓	↑
Olivereau and Lambert, 1981	expt 1: Avoidance learning	↓	↑
	expt 2: Memory	↓	0
	expt 3: Spontaneous activity	↓	↑
Olivereau et al., 1981	EEG slow wave		
	Frequency	↓	↓
	Amplitude	↑	↓
Terry et al., 1969	Maze learning/performance	–	↑

[a] – = not studied; 0 = no effect; ↑ = increase; ↓ = decrease.

Table 7
SUMMARY OF EXPERIMENTAL CHARACTERISTICS: ANIMAL PHYSIOLOGY AND BEHAVIOR

Study		Air ion level (+)	Air ion level (−)	Exposure period (+)	Exposure period (−)
Bachman et al., 1965		5.3×10^5	5.1×10^5	30 min	30 min
Bachman et al., 1966a		1.5×10^6	1.5×10^6	45 min	45 min
Bachman et al., 1966b		1.5×10^6	1.5×10^6	45 min	45 min
Bailey et al., 1983		5.0×10^5	5.0×10^5	18 hr	18 hr
Bailey and Charry, 1984a		5.0×10^5	5.0×10^5	2, 18, 66 hr	2, 18, 66 hr
Bailey and Charry, 1984b		5.0×10^5	5.0×10^5	2, 18, 66 hr	2, 18, 66 hr
Bauer, 1955	expt 1:	6.0×10^3	3.5×10^3	23 hr/day for 31 days	23 hr/day for 31 days
	expt 2:	6.5×10^3	3.0×10^3	20 days	29 days
Diamond et al., 1980		—	1.0×10^5	—	20 days
Dowdall and deMontigny, 1983		1.5×10^6	1.5×10^6	21 days	21 days
Duffee and Koontz, 1965		2.9×10^5	1.4×10^5	Uncertain	
Falkenberg and Kirk, 1977		1.0×10^5	1.0×10^5	—	8 hr
Frey, 1967		NR	NR	—	1 hr/day 13 days
Gilbert, 1973		—	3.0×10^3		100 days
Herrington and Smith, 1935		—	1.2×10^6	—	279 days
			0.6×10^6		
			0.3×10^6		
Jordon and Sokoloff, 1959		—	8.5×10^3	—	3 hr
Lambert and Olivereau, 1980		8.0×10^4	8.0×10^4	40 days	40 days
Krueger et al., 1963	Low	4.5×10^3	2.3×10^3	18 days	18 days
	High	5.1×10^4	5.2×10^4		
Krueger et al., 1966		$4.3—5.0 \times 10^5$	—	22—24	—
Krueger et al., 1968		$4.0—5.0 \times 10^5$	$4.0—5.0 \times 10^5$	4—12 days	4—12 days
Krueger and Kotaka, 1969	High	$3.5—5.0 \times 10^5$	$3.0—5.0 \times 10^5$	12—72 hr	12—72 hr
	Low	$3.0—4.0 \times 10^3$	$2.0—4.0 \times 10^3$		
Lambert et al., 1981		8.0×10^4	8.0×10^4	3 weeks	3 weeks

Table 7 (continued)
SUMMARY OF EXPERIMENTAL CHARACTERISTICS: ANIMAL PHYSIOLOGY AND BEHAVIOR

Study	Air ion level		Exposure period	
	(+)	(−)	(+)	(−)
McDonald et al., 1965	3.5×10^5	3.6×10^5	30 min	30 min
Mizusawa et al., 1981	NR	NR	20 days	20 days
Nazzaro et al., 1967	2.5×10^5	3.1×10^5	8 days	8 days
Olivereau, 1970a	–	6.0×10^5	–	25 min
Olivereau, 1970b	6.0×10^5	–	30 min	–
Olivereau, 1979	7.5×10^5	7.5×10^5	4—6 min	4—6 min
Olivereau et al., 1981	8.0×10^4	8.0×10^4	3 weeks	3 weeks
Olivereau and Lambert 1981	$6.0—6.5 \times 10^5$	$6.0—6.5 \times 10^5$ expt 1:	30 min	30 min
	8.0×10^4	8.0×10^4 expt 2:	3 weeks	3 weeks
	7.5×10^5	7.5×10^5 expt 3:	NR	NR
Terry et al., 1969	–	$7.0 \times 10^6—7.0 \times 10^7$	5 hr	5 hr

b. Learning and Performance

Seven studies (see Table 6) that examined learning and performance in rats show fairly consistent results. In three studies of the effects of ions on avoidance learning,[37-39] exposure to high levels of positive air ions (approximately 10^5 ions/cm^3) was found to interfere with avoidance learning (also called avoidance acquisition), while negative ion exposure at equally high levels tended to enhance it, with one exception.[38] In order for ion exposure to have an effect on avoidance learning, the ions would either have to interfere with the animal's usual ability to learn and perform or alter the animal's degree of sensitivity to the noxious or painful stimulus.

The strength of this consistency is somewhat clouded when the data of the two studies on avoidance learning are closely examined. In the study undertaken by Olivereau and Lambert,[39] effects of negative and positive ions were considered on three indexes of avoidance learning: latency to successful jump, latency to escape, and ability to escape. The data from this study reveal that for only one index, latency to successful jump, did animals treated with negative ions exhibit significant differences in performance (in this case, shorter latencies) when compared with control animals. There were no significant differences between exposed and control animals for this index when positive ions were administered, nor for the other two indexes, regardless of the polarity of the ion levels used.

In the study by Falkenberg and Kirk,[37] animals were grounded through the skull, so that it is likely that either positive or negative current was being conducted directly through the skull to ground. The effects of this do not seem to parallel ordinary in vivo exposure of either animals or humans to air ion concentrations. In addition these investigators did not use a control group, so that the magnitude of the effects cannot be evaluated.

Four other studies reporting findings on learning and performance have yielded mixed results. Data were reported indicating no effects of ion exposure on maze learning in rats and complex and inconsistent effects on performance.[40-42] Data have also been reported indicating that the ability of rats to swim a water maze was improved by exposure to very high levels of negative air ions (on the order of 10^5 to 10^7 ions/cm^3).[43]

c. Serotonin

Three studies, shown in Table 6, reported that exposure to either positive or negative air ions was associated with decreased levels of brain serotonin (5-HT) in rats.[44] This effect has been reported to occur with exposure to ion concentrations of approximately 10^5 ions/cm^3 for periods as short as 12 hr in mice,[45] and as long as 20 days in rats.[44] Two studies have examined 5-HT in several rat brain regions following exposure to 5.0×10^5 ions/cm^3 (positive and negative unipolar) for periods up to 66 hr and found no effects. Serotonin is a neurotransmitter implicated in the regulation of sleep, sexual activity, and aggression in both animals and humans. The data on brain serotonin indicate that pathological conditions or disordered behavior occur when serotonin levels are depleted 50% or more. For example, depletion of this magnitude has been associated with suppression of mating behavior in females,[46] hyperphagia (i.e., increased food consumption),[47] increased convulsive activity,[48] inhibition of open field activity, and impaired acquisition of avoidance tasks.[49] It has also been observed that animals with depleted levels of serotonin in the brain are more behaviorally agressive, anxious, and fearful.[50,51]

The behavioral implications of the three studies which show a relationship between brain serotonin levels and ion exposure are unclear in light of these findings. Two of these studies reported that exposure to either positive or negative air ions was associated with a decrease in brain serotonin levels of less than 20%,[45,52] while the third reported reductions on the order of 60%.[44] Thus against the background of a significant body of serotonin-behavior research not related to ions, only one of these studies suggests the possibility that ions could influence behavior by altering brain serotonin.

The behavioral implications of noninduced changes in serotonin are also uncertain in view of the results of other studies which indicate that positive and negative air ions have opposite effects on behaviors which are presumably mediated by 5-HT. One study, for example, reported that avoidance learning in rats was facilitated by exposure to negative ions and impaired by positive ion exposure.[37] Such data would seem to contradict the findings that brain serotonin levels have been shown to decrease during exposure to either positive or negative ions.

Overall, the data currently available to assess the accuracy of the hypothesized link between ions and behavior through serotonin is limited. According to this hypothesis, avoidance learning should be impaired with exposure to either positive or negative ions, but the data are not supportive, and in fact are contrary. Given the data showing no effects of ion exposure on brain 5-HT, it appears that the observed effects of ions on brain 5-HT were either extremely weak or produced by artifact.

Looking at peripheral 5-HT in mice,[53-55] the data are mixed for effects of negative ions, but agree that positive ion exposure for periods ranging from 4 to 24 days were associated with elevated blood 5-HT levels.

d. Pain Sensitivity

Two studies shown in Table 6 examined the effects of negative and positive ion exposure on sensitivity to pain in rats.[56,57] These studies indicate that positive ion concentrations of 6.0×10^4 ions/cm^3 for 25 to 30 min increased pain sensitivity, while equivalent negative ion levels decreased it. Pain sensitivity in these studies was defined operationally as the latency period between being placed on a hot plate (heated at 64°C) and subsequent licking of paws by the exposed animal. The pain being tested therefore was pain due to burns. A longer latency was interpreted as lower pain sensitivity and a shorter latency as the opposite. Caution in generalizing from these findings is necessary because to date no replication has occurred.

e. Physiological Arousal and Activity

The pattern of findings for effects of ions on physiological arousal and activity are less straightforward than those previously discussed, and in some cases the reported findings are contradictory.

Three studies shown in Table 6 investigated the effects of ions on physiological arousal in rats; one considered two indicators of autonomic activity — respiration and heart rate — and two others focused on indicators of CNS activity.[39,58,63] It appears that exposure to high concentrations of ions of either polarity (approximately 10^5 ions/cm^3) for 30 min increased peripheral autonomic arousal. By contrast, CNS arousal was decreased by positive ion exposure, as indicated by decreased EEG frequency and increased amplitude in hippocampal and cortical areas. CNS arousal was seen to be increased by negative ion exposure, as evidenced by the occurrence of the opposite effects on EEG activity.

While there is no definitive explanation for the different reactions to ionized air exhibited by the autonomic and central nervous systems, such differences are not uncommon. Investigations of the significance of such differences suggest that these systems may act in opposition in order to maintain physiological equilibrium within an organism.[59]

The data relating ions and activity are contradictory (see Table 6). Three studies examined the effects of ions on spontaneous activity in rats. One reported that positive ions were associated with increased activity,[60] while negative ions were related to activity in a nonlinear fashion (activity first increased and then decreased as a function of increased exposure levels). In the other two studies different effects were observed — positive ions decreased activity and negative ions increased it.[39,61] The ion levels used were different for these studies (1.5×10^5 ions/cm^3 for 30 min for the first study; 7.5×10^5 ions/cm^3 for 6 min for the other two studies), but the differences do not appear large enough to justify these opposite outcomes. Based on these studies, the effects of ions on activity are unresolved, though limited evidence suggests that positive ions might be associated with decreased activity and negative ions with increased activity. On the other hand, data has been reported indicating no effects of 5.0×10^5 (+) or (−) ions/cm^3 on spontaneous motor activity for exposure periods of 2, 18, and 66 hr.[34] Taken together, the data for behavioral effects of ions in test animals suggest either weak effects, or no effects.

2. *Critique of Experimental Design and Documentation*

The important characteristics of the experimental environment for the studies of animal physiology and behavior are outlined in Table 8. Of 28 studies, 6 reported electric field strength for the experiments;[33,34,39,56,57,62] 3 studies indicated current density.[33,34,62] Ion mobility was reported in nearly half of the studies, as were temperature and humidity. Barometric pressure was uniformly not reported.

In 12 of the 28 studies, the air delivery system used was inadequately described, and in few cases were gaseous contaminant levels in the test environment estimated (see Table 9). Consequently, the possibility that the presence of aerosols or gaseous contaminants affected results cannot be dismissed, and it cannot be assumed that small air ions were causative agents in the observed effects.

As Table 10 indicates, some of the studies of animal physiology and behavior were designed correctly. In most cases (21 of 28) animals tested were grounded during experiments, and the results were subjected to appropriate statistical analysis.

Among these 28 studies are attempts to replicate previous results (Table 10) in the areas of learning, neurochemistry and behavioral activity. Two studies that examined the results of negative ion exposure on conditioned emotional response (CER) represent a partial replication attempt.[35,36] Each study found that negative ions decreased the CER in rats, but Nazzaro et al. were able to observe this in only half of their test animals.[36]

Other replication attempts concern performance and learning. Studies of performance in rats exposed to negative ions showed improvement with negative ions.[42,43] Individual differences (age and sex) in response to air ions were also reported. Exposure to positive ions was reported to impair avoidance learning, while negative ions were found to improve it.[37,39]

In the area of neurochemistry, three studies reported that negative ions reduced brain serotonin levels in animals.[44] They are not directly comparable, however, because different species were used as test subjects (mice vs. rats), ion levels were different (10^3 vs. 10^5 ions/cm^3) and exposure periods varied (hours vs. days). Two studies were replicated showing no effects of either positive or negative ions on brain 5-HT.[33,62]

Finally, attempts at replication were made for three studies (Table 10) in the area of behavioral activity.[34,39,61] In the study by Bailey and Charry, no effects of either positive or negative ions were reported on spontaneous motor activity.[34] In the studies by the Olivereau group, spontaneous activity decreased in the presence of positive air ions and increased with negative air ions. Based on these data, effects on behavioral activity are unresolved.

3. *Results of Minimally Acceptable Studies*

Out of the literature reviewed, there are seven studies on physiological and behavioral responses in animals that meet the criteria for minimally acceptable scientific rigor (see Tables 6 to 10). Each is reviewed briefly below.

Bachman et al. found that exposure of rats for 30 min to either positive or negative air ions (5.0×10^5 ions/cm^3) counteracted the effects of anesthesia (44 mg/kg sodium pentobarbital, i.p.) by increasing heart rate.[63] Respiration was tested in this manner but was not significantly affected. Control animals (injected with anesthetic, but not exposed to ions) showed a decline in heart beat frequency by 59 beats per minute. By comparison, the

Table 8
DC ELECTRICAL ENVIRONMENTAL CHARACTERISTICS: ANIMAL PHYSIOLOGY AND BEHAVIOR

Study	Field (V/cm)	Current density (A/cm^2)	Mobility (cm^2/V-s)	Ion exposure system
Bachman et al., 1965	NR[a]	NR	1.0	Kr 85 ion producing source (beta emitter), ions transported by airflow and electric field, 250 V to 2000 V used for charge separation, ion concentration calculated from current density
Bachman et al., 1966a	NR	NR	1.0	Same as Bachman et al., 1965
Bachman et al., 1966b	NR	NR	1.0	Same as Bachman et al., 1965
Bailey et al., 1983	30	4.0×10^{12}	1.4—1.8	Exposure to ions by corona discharge and measurement by volumetric ion counter, ion transport by laminar flow and electric field
Bailey and Charry, 1984a	30	4.0×10^{12}	1.4—1.8	Same as Bailey et al., 1983
Bailey and Charry, 1984b	30	4.0×10^{12}	1.4—1.8	Same as Bailey et al., 1983
Bauer, 1955	NR	NR	NR	Exposure to ionized air using polonium 210 with/ss biasing electrodes
Diamond et al., 1980	NR	NR	1.0—2.0	Ions generated by corona discharge and measured with parallel plate ion counter
Dowdall and de Montigny, 1983	NR	NR	NR	NR
Duffee and Koontz, 1965	NR	NR	0.02	Kr 85 source for ion generation, applied voltage of 6 kV
Falkenberg and Kirk, 1977	NR	NR	NR	NR
Frey, 1967	NR	NR	NR	Titanium foil and tritium gas ionizer, ions dispersed by air
Gilbert, 1973	NR	NR	NR	Titanium foil and tritium gas ionizer, using biased 220 V ss screen. Ions measured 2 in. above cage floor with a collecting tube
Herrington and Smith, 1935	NR	NR	1.0—1.5 0.0056—1.0 0.0007—0.0056	Ions generated by point corona discharge and measured with volumetric counter
Jordon and Sokoloff, 1959	NR	NR	NR	Ions generated by polonium foil and blown through wooden exposure chambers by oscillating fans and measured with ion counter
Krueger, et al., 1963	NR	NR	1—2	Ions generated either by krypton-84 or tritium with focusing plexiglass electrodes, and measured with ion counter
Krueger, et al., 1966	NR	NR	NR	Ions generated by tritium with focusing plexiglass electrodes, measurement same as Krueger et al., 1963
Krueger, et al., 1968	NR	NR	1.0—2.0	Ions generated by tritium absorbed on zirconium. Charge separation using rectifying circuit of 950 V. Ion measurement same as Krueger et al., 1963
Krueger and Kotaka, 1969	NR	NR	1.0—2.0	Same as Krueger et al., 1968
Lambert and Olivereau, 1980	NR	NR	NR	Ions generated by point corona discharge and measured by ionometer
Lambert, et al., 1981	NR	NR	NR	Same as in Olivereau et al., 1981

McDonald et al., 1965	NR	NR	1.0	Same as Bachman et al., 1965
Mizusawa et al., 1981	NR	NR	NR	NR
Nazzaro et al., 1967	NR	NR	1.2	Ions generated by "tritium gas sealed in zirconium on stainless steel" Ions measured by parallel plate collector
Olivereau, 1970a	10	NR	NR	NR
Olivereau, 1970b	10	NR	NR	NR
Olivereau, 1979	NR	NR	NR	NR
Olivereau et al., 1981	NR	NR	NR	Ions generated by point corona discharge and measured at ground by ionometer
Olivereau and Lambert, 1981	3	NR	NR	Same as Olivereau et al., 1981
Terry et al., 1969	NR	NR	NR	Ions produced by beta emission and dispersed by fan

[a] NR = not reported

Table 9
MICRO-CLIMATE ENVIRONMENTAL CHARACTERISTICS: ANIMAL PHYSIOLOGY AND BEHAVIOR

Study	T (C)	RH (%)	BP	O_3 (ppm)	NO_x (ppm)	Air delivery system
Bachman et al., 1965	NR[a]	NR	NR	NR	NR	Gelman membrane filter 99% particle retention >0.05μm, airflow 80 liters/m; ion mobility from estimated filters, not measured directly
Bachman et al., 1966a	17	20—65	NR	NR	NR	Same as Bachman et al., 1965
Bachman et al., 1966b	NR	NR	NR	NR	NR	Same as Bachman et al., 1965
Bailey et al., 1983	23 ± 5	45 ± 5	765.5	<0.01	<0.1	Air treated (heated or cooled, humidified), with large particle filter, activated carbon, 99.97%, 0.3μ HEPA filter, electronic air cleaner, at 0.6 m^3/min
Bailey and Charry, 1984a	23 ± 1	45 ± 5	765.5	<0.01	<0.1	Same as Bailey et al., 1983
Bailey and Charry, 1984b	23 ± 1	45 ±	765.5	<0.01	<0.1	Same as Bailey et al., 1983
Bauer, 1955	11—23	NR	NR	NR	NR	Enclosed wooden chambers with micro-climate controls, airflow through chambers at 0.08—0.11 cfm
Diamond et al., 1980	NR	NR	NR	NR	NR	Filtered air (filter size not specified) blown by fan into exposure chambers
Dowdall and de Montigny, 1983	NR	NR	NR	NR	NR	NR
Duffee and Koontz, 1965	21—23	50	NR	NR	NR	Airflow through environmentally controlled chambers 162 ft/min
Falkenberg and Kirk, 1977	24	45	NR	NR	NR	NR
Frey, 1967	20—23	45—65	NR	NR	NR	Negative pressure system drawn by muffin fan with electrostatic and paper air filtration having particle retention capability of >0.8μm
Gilbert, 1973	20 ± 2	37 ± 3	NR	NR	NR	Airflow at 17.5 ℓ/min
Herrington and Smith, 1935	NR	NR	NR	est.[b]	NR	Air-conditioned exposure chambers, temperature and humidity records were kept but were not reported
Jordon and Sokoloff, 1959	24	NR	NR	NR	NR	Ions generated by polonium foil and blown through wooden exposure chambers by oscillating fan
Krueger et al., 1963	24	56 ± 6	NR	NR	NR	Ions generated either by krypton-84 or tritium with plexiglass focusing electrodes, metal cages with clean air delivered at 122 cfm and flow through cage at 103 cfm. All exposure within sealed cubicles
Krueger et al., 1966	26	42	NR	NR	NR	Air delivered at rate of 623 ℓ/min through electrostatic pre-

						cipitator, activated carbon and air conditioner. Animal containment section consisted of grounded metal floor with plexiglass walls
Krueger et al., 1968	26	45	NR	NR	est.	Electrostatic, activated carbon, and absolute filters with airflow through each exposure chamber of 0.178 m^3/min
Krueger and Kotaka, 1969	26 ± 1	45 ± 2	NR	NR	est.	Same as Krueger et al., 1968
Lambert and Olivereau, 1980	22	60—65	NR	NR	NR	Cages housed in ventilated room
Lambert et al., 1981	22	60—65	NR	<0.01	NR	Same as in Olivereau et al., 1981
McDonald et al., 1965	NR	NR	NR	NR	NR	Same as Bachman, et al., 1965
Mizusawa et al., 1981	NR	NR	NR	NR	NR	NR
Nazzaro et al., 1967	23	80	NR	NR	NR	NR
Olivereau, 1970a	18	NR	NR	est.	est.	NR
Olivereau, 1970b	18	NR	NR	est.	est.	NR
Olivereau, 1979	NR	NR	NR	NR	NR	NR
Olivereau et al., 1981	22—23	60—65	NR	<0.01	NR	Electrostatic precipitation of large particles ventilation mentioned but unspecified, cages cleaned daily, ozone measurement method not described
Olivereau and Lambert, 1981	23	60—65	NR	est.	est.	Vacuum device used to eliminate ozone and nitrogen oxides, though no direct measures were taken
Terry et al., 1969	22—25	NR	NR	NR	NR	NR

[a] NR = not reported
[b] est. = estimated.

Table 10
EXPERIMENTAL DESIGN, PROCEDURES AND STATISTICAL TESTS: ANIMAL PHYSIOLOGY AND BEHAVIOR

Study	Design	Grounding	Statistics	Replication	Other experimental issues/problems
Bachman et al., 1965	Factorial	Yes	F test	No	Errors in data presented in tables (lack of fit with the text), basis of calculating ion density from ion current is unclear
Bachman et al., 1966a	Factorial	Yes	F test	No	Basis for ion calculation unclear
Bachman et al., 1966b	Factorial	Yes	No	No	Same as for Bachman et al., 1966a
Bailey et al., 1983	Independent three group Double-blind	Yes	t test	Yes	
Bailey and Charry, 1984a	Independent three group Counter-balanced exposure Double-blind	Yes	t test	Yes	
Bailey and Charry, 1984b	Independent three group Counter-balanced exposure Double-blind	Yes	t test	Yes	
Bauer, 1955	Factorial	NR[a]	F test	No	Wide variation in temperature used to induce nest building a possible confounding variable for detecting ion effects; materials used in nest building studies with varying dielectric properties have potential to significantly alter ion concentration with resultant dosimetry problems
Diamond et al., 1980	Factorial	Yes	t test	Partial	Animals exposed in cubicles containing sawdust bottoms, question about dust causing presence of large ions or aerosols, changing sawdust 3 times/week allows buildup of animal waste and protein on walls which might threaten the integrity of the ion concentration
Dowdall and deMontigny, 1983	Independent three	NR	NR	No	Report available only in abstract form and

	group				therefore difficult to assess methodology
Duffee and Koontz, 1965	Factorial Counterbalancing	Yes	F test	Partial	Learning and performance not well defined, errors in reporting the data, tables not explained
Falkenberg and Kirk, 1977	Independent two group	Yes	Fisher exact test	Partial	Lack of control group (ambient conditions) makes data difficult to interpret
Frey, 1967	Independent three group	Yes	Mann-Whitney U	Partial	
Gilbert, 1973	Factorial	NR	X^2	Partial	Use of one-way avoidance conditioning with animals following ion exposure and prior to decapitation and 5-HT analysis confounds data since effect of conditioning on brain 5-HT is not reported; emotionality assessment by reactivity to handling is not usually a reliable measure, unless used in conjunction with other measures
Herrington and Smith, 1935	Independent four group	NR	Critical ratio	No	Effects on activity were observed between 175—300 days, but not prior to 175 days
Jordan and Sokoloff, 1959	Independent four group	NR	NR	No	Learning test occurred 20 min after animals were exposed to ions; since decay rate of ion effects is not established, this procedure contains inherent artifacts; inadequate description of ion measurement procedures
Krueger et al., 1963	Independent three group	Yes	t test	Partial	Cage design and use of plastic film under cages present likely dosimetry problems, although data given suggest uniform ion concentrations among cages; lack of pre- and post-ion measurements does not allow accurate estimate of animal exposure
Krueger et al., 1966	Independent two group	Yes	NR	Partial	Expanded metal flooring may be stressful; lack of statistical tests with large error variance makes results difficult to interpret; use of grounded floor surrounded by plexiglass raises questions of dosimetry and micro-shock due to potential differences
Krueger et al., 1968	Repeated measures	Yes	NR	Partial	Lack of statistical tests and large variance make significance of differences difficult, if not impossible, to interpret

Table 10 (continued)
EXPERIMENTAL DESIGN, PROCEDURES AND STATISTICAL TESTS: ANIMAL PHYSIOLOGY AND BEHAVIOR

Study	Design	Grounding	Statistics	Replication	Other experimental issues/problems
Krueger and Kotaka, 1969	Factorial	Yes	No	Partial	Report of 24-hr acclimatization period unclear as to location and whether ions were present during this period
Lambert and Olivereau, 1980	Independent three group Double-blind	Yes	Non-para F test Mann-Whitney U		Use of exposure cages with grounded walls and floor makes dosimetry questionable; inadequate description of ion measurement techniques, cages may have collected all ion current generated
Lambert et al., 1981	Independent three group	Yes	t test	No	
McDonald et al., 1965	Factorial	Yes	F test	No	Animals exposed to ion conditions without inhalation of ions; limited results were compared with Bachman et al., 1966, on basis of inhalation as a key variable; lack of effects attributed to lack of inhalation
Mizusawa et al., 1981	Independent three group	NR	NR	No	Almost total lack of information on ion exposure, methods, or procedures makes data difficult to evaluate; confusing explication of data, reference to significane levels, but no indication of statistics used
Nazzaro et al., 1967	Repeated measures	No	NR	Partial	Rats held at floating potential which does not ensure consistent exposure to ions; estimation of ion concentration from ion current with no data showing the relationship between the two measures; no tests for differential transfer (+) to (−) or (−) to (+), and no description of detoxification period between exposure to different unipolar conditions raised questions about artifact due to contamination or carryover effects
Olivereau, 1970a	Independent two group	Yes	t test	No	Data reported significant only at $P < 0.10$, normally not considered adequate for avoid-

					ing statistical error
Olivereau, 1970b	Independent two group	Yes	t test	No	
Olivereau, 1979	Independent three group	Yes	X^2	Yes	
Olivereau and Lambert, 1981	Independent three group, exp 1, 2 Repeated measures expt 3	Yes	t test X^2 Mann-Whitney U	Partial	Learning test occurred 15—45 min after animals were exposed to ions; since decay rate of ion effects is not established, this procedure contains inherent artifacts; in expt 3, investigators reported that rats were ungrounded, and therefore it is not possible to determine the extent to which rats were exposed to air ions
Olivereau et al., 1981	Independent three group	Yes	t test	No	
Terry et al., 1969	Randomized block	NR	F test	Partial	Mapping of ion environment at 20 locations reported, but locations and degree of ion uniformity unspecified; use of ion current to determine ion concentration with no details on how this was accomplished; statistical analysis and presentation confusing, resulting in uncertainty about data reported

[a] NR = not reported

treatment group (rats injected with anesthetic and exposed to ions) showed an average decline in heartbeat of 30 beats per minute. Exposure was observed, therefore, to inhibit the usual effects of anesthesia.

Several studies reported increased neurophysiological arousal in EEG recordings in rats exposed for 3 weeks to negative ion concentrations of 8.0×10^4 ions/cm^3.[39,58] The opposite response was observed for positive ion exposure. In one case, positive ions resulted in significantly lowered CNS arousal relative to controls, whereas negative ions had no effect.[58] In the other case positive ions significantly decreased arousal in the brain relative to controls, and negative ions significantly increased arousal relative to controls.[39]

Decreases in the level of brain serotonin (5-HT) and cyclic AMP have been reported to accompany ion exposure.[44] Investigators exposed rat pups to negative air ions (1.0×10^5 ions/cm^3) for 20 days. However, changes in these neurochemicals in more than one brain region occurred only when negative ions were applied to rat pups in a multifamily "enriched" environment (5-HT levels decreased by 45 to 61% in the somatosensory and occipital cortex; cyclic AMP was decreased by 35 to 45% in both regions). Rats exposed to negative ions in a unifamily "impoverished" environment showed no change in serotonin in any of the brain regions studied (somatosensory or occipital cortex) and a change in cyclic AMP was observed in the somatosensory cortex only. Since the effects of social environmental factors were so pronounced in this study, the biological significance of effects of ions is unclear.

A reasonably well-designed and controlled study showed that mice exposed to negative ions (6.5×10^5 ions/cm^3 for 30 min) learned to escape from a noxious stimulus significantly faster than controls.[39] Positive ion exposure did not significantly affect avoidance learning. In the same study positively ionized (8.0×10^4 ions/cm^3 for 3 weeks) animals showed significantly less memory retention than controls.

In studies examining effects of ions on spontaneous activity, Olivereau and Lambert exposed rats to positive or negative ions (7.5×10^5 ions/cm^3 for 20 min) and showed significant differences in activity compared to control animals with positive ions decreasing and negative ions increasing it.[39] In a second study,[64] positive ions (1.5×10^6 ions/cm^3 for 45 min) affected activity of rats in an inconsistent manner, and negative ions decreased it.

Finally, the most recent studies shown in Table 6 resulted in entirely negative results.[33,34,62] The control and characterization of the exposure environment in these studies, as illustrated in Tables 6 to 10 were extremely rigorous. In these experiments the effects of positive and negative unipolar ion conditions (5.0×10^5 ions/cm^3) were studied in rats over three different time periods — short-term exposure (2 hr), moderate exposure (18 hr), and long-term exposure (66 hr). Included as dependent measures were turnover (rate of disappearance) of three neurotransmitter systems — serotonergic, noradrenergic and dopaminergic — in five different brain regions. Spontaneous motor activity as well as food and water consumption were examined. No effects of ion exposure occurred on any of the neurotransmitter systems in any of the brain regions examined; also, no effects on spontaneous motor activity and no effects on food or water consumption occurred. These data represent the most reliable and valid to date in rats exposed individually up to 3 days, as they emerged from well-controlled, internally replicated studies.

The effects reviewed in this section can be summarized as follows: first, air ions of both polarities exerted cardiovascular and respiratory effects on anesthetized animals, where ion exposure returned heart rate and respiration to normal, equaling that recorded for unanesthetized animals. The exposure levels for these animals were 10^5 ions/cm^3 for 3 to 20 days. Second, there appeared to be some effect of air ion exposure on motor activity at levels of 10^5 and 10^6 ions/cm^3 when animals were exposed for at least 30 min. Third, contradictory data exist for effects on brain serotonin (5-HT) and activity level at ion concentrations of 10^5 or 10^6 ions/cm^3. The best-controlled studies in this area were in rats and reported no effects of either positive or negative air ions on neurochemical activity including serotonin, and no effects on spontaneous motor activity.

C. Human Biomedical Responses

A total of 18 clinical or clinically related studies were reviewed here. The results of these studies are summarized in Table 11. The experimental characteristics and procedures are presented in Tables 12, 13, 14 and 15, respectively.

1. Reported Results: Aggregate Data

The studies reviewed on the effects of air ions on various human biological responses implicated in clinical disorders indicate that ions exerted no effects on pulmonary function, pulse, blood pressure, respiration rate, or emphysema. On the other hand, a number of studies revealed ion effects on human pathological conditions including pain due to burns, respiratory disorders, and various common symptoms such as headache, sore throat, and upper respiratory congestion.

Where effects were observed, it is the general pattern from the results of these clinical studies that negative ions ameliorated health disorders, while positive ions aggravated them. A number of studies, however, reported no effects. The contradictory data may be due in part to differences in ion exposure level and exposure period among studies; they also may be due to any of a number of design errors and experimental artifacts.

Three clinical studies examined the effects of negative air ions on postoperative discomfort and on pain associated with burns in humans;[65-67] all agree that negative air ions had an analgesic effect on these conditions (see Table 11). The effects of positive ions were not examined in these studies. Conclusive generalizations on the basis of these works, however, are not possible because of the incomplete reporting and confounding procedures in these studies, as well as a possible placebo effect (see Tables 13, 14, and 15).

Seven studies reported contradictory results from exposure of people with various respiratory disorders to air ions.[68-74] Two of these reported that negative air ions eased bronchial asthma, respiratory spasticity, and asthmatic bronchial spasticity. A close look reveals information that makes interpretation of the data difficult. In the study by Jones et al.[70] it was reported that of seven patients tested, only three showed statistically significant improvement in respiratory function as measured by Peak Expiratory Flow Rate (PEFR) during both morning and afternoon readings. The remaining four subjects showed either improvement only during morning readings, no improvement, or a decrement in PEFR. Subjects in this study were exposed to unreported ion concentrations over an 8-week period. One other study examining PEFR reported no effects due to ion exposure.[74] Research (reported in two different journals) on patients with bronchial asthma indicated lung function effects (increased forced expiratory volume) with exposure to either positive or negative air ions,[66,72] but no effects on histamine threshold. Motley and Yanda[71] examined effects of ions on respiratory activity and concluded that neither positive nor negative ion exposure at very high levels affected the manifestation of symptoms in people with emphysema. In the area of respiration, there are contradictory data among the studies and too many potential experimental artifacts in the work which has been completed to draw any conclusions.

Perhaps the most consistent of the studies shown in Table 11 are those that examined the effects of ions on peripheral serotonin (5-HT) and 5-hydroxy-indoleacetic acid (5-HIAA), a primary 5-HT metabolite. These studies, all conducted by the Sulman group in Israel,[76,78,80-83] reported that exposure to naturally occurring positive air ion concentrations (due to Sharav weather conditions) of 10^3 ions/cm^3 or to ion concentrations under laboratory conditions of 10^5 ions/cm^3 was associated with increased urinary 5-HT and 5-HIAA for weather-sensitive persons, suggesting increased metabolism of 5-HT. Moreover, two of the Sulman group studies also indicated that exposure to negative air ions was associated with decreased urinary 5-HT and 5-HIAA.

A study of hospital patients, summarized in Table 11, indicated that exposure to either positive or negative ions at levels of approximatley 10^4 ions/cm^4 produced symptoms such

Table 11
SUMMARY OF EXPERIMENTAL RESULTS: HUMAN BIOMEDICAL RESPONSES

Study	Biomedical response	Air ion effects[a] (+)	Air ion effects[a] (−)
Albrechtsen et al., 1979	Bronchial asthma		
	Lung function	↑	↑
	Histamine threshold	0	0
	Air quality rating	0	0
Dantzler et al., 1983	Pulmonary function	0	0
	Pulse	0	0
	Blood pressure	0	0
	5-HIAA	0	0
	Mood	0	0
David et al., 1960	Burn pain	–	↓
David et al., 1962	Postoperative discomfort	–	↓
Hawkins, 1981	Headache	–	0↓
	Nausea	–	0↓
	Dizziness	–	0↓
Jones et al., 1976	Bronchial asthma	–	0↓
Minehart et al., 1961	Postoperative discomfort	–	↓
Motley and Yanda, 1966	Emphysema	0	0
Osterballe et al., 1979	Bronchial asthma		
	Lung function	↑	↑
	Histamine threshold	0	0
	Air quality rating	0	0
Palti et al., 1966	Respiratory spasticity	↑	↓
	Asthmatic bronchial spasticity	↑	↓
Sulman et al., 1970	Irritation syndrome		
	5-HT urine	↑	–
	5-HIAA urine	↑	–
	Exhaustion syndrome		
	NE, EPI urine	↓	–
	17-KS, 17-OH urine	↓	–
	Hyperthyroidism		
	5-HT urine	↑	–
	NE, EPI urine	↑	–
	17-KS, 17-OH urine	↑	–
	Histamine urine	↑	–
Sulman, 1971	Serotonin-migraine	↑	–
Sulman et al., 1973	Exhaustion syndrome		
	NE, EPI urine	0	–
Sulman et al., 1974	Irritation syndrome		
	5-HT urine	↑	↓
	5-HIAA urine	↑	↓
Sulman et al., 1975	Irritation syndrome		
	5-HT	↑	↓
	5-HIAA urine	↑	↓
Sulman et al., 1978	Body weight	–	0
	Blood pressure	–	0
	Pulse	–	0
	Respiratory rate	–	0
	Oral temperature	–	0
	Urinary pH	–	0
	Albumen	–	0
	Glucose	–	0
	Ketones	–	0
	Bilirubin	–	0

Table 11 (continued)
SUMMARY OF EXPERIMENTAL RESULTS: HUMAN BIOMEDICAL RESPONSES

Study	Biomedical response	Air ion effects[a] (+)	Air ion effects[a] (−)
	Occult blood	–	0
	Blood cells, red and white	–	0
	ECG	–	0
	EEG		
	Alpha	–	↑
	Synchronization	–	↑
	5-HT, 5-HIAA		
	Weather-sensitive	–	↓
	Normal	–	0
	NE		
	Weather-sensitive	–	↑
	Normal	–	0
Wagner et al., 1983	Bronchial asthma		
	Peak flow rate	0	–
Winsor and Beckett, 1958	Headache	↑	0
	Dizziness	↑	0
	Sore throat	↑	0
	Smarting eyes	0	↓
	Congested throat	0	↓

[a] Key: – = not studied; 0 = no effects; ↑ = increase; ↓ = decrease.

as headache, dizziness, and sore throat.[75] Each of these symptoms was rated as mild by the test subjects who experienced them.

Data also have been reported indicating that longer-term exposure (2 months) to negative air ions on the order of 10^4 ions/cm^3 had no effect on a variety of metabolic functions, although changes in alpha wave activity indicative of decreased neurophysiological arousal seemed to occur.[76] These data are in general agreement with others that reported effects of air ions on brain wave activity. Generally, negative ion exposure was associated with decreased alpha frequency and increased amplitude in humans.[77] Similar data for slow-wave activity have been reported for animals.[39,58]

2. *Critique of Experimental Design and Documentation*

As is typical of most of the literature, reporting of experimental conditions for the human clinical studies is inadequate. None of the 18 studies reported either field strength or current density; only one study reported ion mobilities operative during experiments, but did not describe the technique employed to measure mobility (see Table 13).

About half of the studies provided data on the temperature and humidity during experiments (see Table 14). Levels of ozone (O_3) and oxides of nitrogen (NO_x) present during testing were reported by only 2 of the 17 studies.[76,78] This is a significant omission since many of the symptoms reported to be associated with ion exposure also can be caused by direct exposure to ozone or oxides of nitrogen such as nitrogen dioxide (NO_2). For example, one study reported that exposure to positive or negative air ions induced various combinations of headache, dizziness, and irritation of throat and eyes. These symptoms also occur as the result of exposure to ozone or NO_2.[79]

Description of the air delivery system used in these studies is uniformly inadequate. No description of the air delivery system was provided by 9 of the 18 studies. The other studies

Table 12
SUMMARY OF EXPERIMENTAL CHARACTERISTICS: HUMAN BIOMEDICAL RESPONSES

Study	Air ion dosages (ions/cm^3)		Exposure period	
	(+)	(−)	(+)	(−)
Albrechtsen et al., 1979	3.0×10^4	3.2×10^4	15 min	15 min
Dantzler et al., 1983	1.0×10^5	1.0×10^5	6 hr	6 hr
David et al., 1960	–	$1.8—7.0 \times 10^3$	–	1—3 times/day for 25 min
David et al., 1962	–	NR[a]	–	NR
Hawkins, 1981				
Site 1	0.5×10^2	2.5×10^3	–	
Site 2	1.3×10^2	2.0×10^3	–	4—6 weeks
Site 3	1.0×10^2	3.5×10^3	–	8 hr/day
Jones et al., 1976	–	NR	–	Each night for 8 weeks
Minehart et al., 1961	–	NR	–	NR
Motley and Yanda, 1966	5.0×10^5	5.0×10^5	7—12 hr/week for 2 weeks	
Osterballe et al., 1979	3.0×10^4	3.2×10^4	15 min	15 min
Palti et al., 1966	10^4	10^4	28 hr	
Sulman et al., 1970	NR	NR	151 days	–
Sulman et al., 1971	NR	NR	50—180 days	–
Sulman et al., 1973	NR	NR	50—180 days	–
Sulman et al., 1974	3.5×10^5	3.5×10^5	30—60 min	60 min
Sulman et al., 1975	3.5×10^5	3.5×10^5	1—12 hr	1—12 hr for 1 day
Sulman et al., 1978	–	1.0×10^4	–	16 hr/day for 2 months
Wagner et al., 1983	$2.3—3.0 \times 10^4$	–	24 hr	–
Winsor and Beckett, 1958	3.5×10^4	2.6×10^4	15—20 min	15—20 min

[a] NR = not reported.

did not describe the system sufficiently well to be sure that only small air ions were present of that air delivered to test subjects was clean and free of contaminants (see Table 14). As a result, it is not certain for any of these studies whether the results obtained should be attributed to air ions or to some other factor.

The concentration of ions used was reported sporadically in this group of studies. As a result, it is difficult to identify from the sum of this clinical research any threshold level of ion exposure above which effects generally could be expected. This is particularly a problem with respect to three studies examining effects of negative ions on pain from burns and postoperative discomfort.[65-67] All three studies agree that negative ions reduced pain, but only one reported the ion level used.

Table 15 summarizes other experimental characteristics of the 18 clinical studies involving biomedical responses in humans. It is evident that the experimental designs vary widely. Taken together, this group of studies is the least well designed and controlled of the four groups of studies reviewed here. In almost all cases description is confusing. Data were often presented in a fashion that does not follow the usual conventions for reporting scientific research. In only half of the studies were statistical tests used. Moreover, there are numerous experimental problems present in most all of these studies which reduce the usefulness of the reported findings.

There are a number of problems common to nearly all of these studies. In only 2 of 18

Table 13
DC ELECTRICAL ENVIRONMENTAL CHARACTERISTICS: HUMAN BIOMEDICAL RESPONSES

Study	Field (V/cm)	Current density (A/cm²)	Mobility (cm²/V-s)	Ion exposure system
Albrechtsen et al., 1979	NR[a]	NR	0.8	Ions generated by alpha emitter with biased grid for controlling polarity
Dantzler et al., 1983	NR	NR	NR	Negative or positive ions generated by corona and measured with volumetric collector at face level, subjects grounded and located across from generator
David et al., 1960	NR	NR	NR	Ions generated by radioactivity and corona, otherwise NR
David et al., 1962	NR	NR	NR	Wire (0.0635 mm radius) corona discharge with aluminum plate collecting electrodes, 6 kV applied to produced corona, ions counted by neon light blinking rate, ozone levels not measured above background
Hawkins, 1981	NR	NR	1—2	Ions generated by corona discharge and dispersed by fan
Jones et al., 1976	NR	NR	NR	Ions produced by "electric-field type" generator, otherwise NR
Minehart et al., 1961	NR	NR	NR	Same as David et al., 1962
Motley and Yanda, 1966	NR	NR	NR	Ions generated by unspecified method and measured with ion collector and electrometer. Method for obtaining ion volume not described
Osterballe et al., 1979	NR	NR	0.8	Ions generated by alpha emitter into airstream and directed at subject, concentration and mobility measured by "cylindrical condensor"
Palti et al., 1966	NR	NR	NR	Ions generated by unspecified method, subject exposure calculated
Sulman et al., 1970	NR	NR	NR	Ions created by weather front known as Sharav in Israel
Sulman et al., 1971	NR	NR	NR	Same as Sulman et al., 1970
Sulman et al., 1973	NR	NR	NR	Same as Sulman et al., 1970

Table 13 (continued)
DC ELECTRICAL ENVIRONMENTAL CHARACTERISTICS: HUMAN BIOMEDICAL RESPONSES

Study	Field (V/cm)	Current density (A/cm^2)	Mobility (cm^2/V-s)	Ion exposure system
Sulman et al., 1974	NR	NR	NR	Ions generated by corona discharge using 2 kV applied to tungsten-iridium wire; ions diffused by fan; output rated at 3.5×10^3 ions/cm^3 sec, ion volume counted
Sulman et al., 1975	NR	NR	NR	Same as Sulman et al., 1974, except for applied voltage of 6 kV
Sulman et al., 1978	NR	NR	NR	Ions generated by point corona discharge with applied voltage of 5 kV; output rated at 2.5×10^{11} ions/s/mm^3, with ion density at 1—2 m of 2.5×10^5 ions/cm^3/s
Wagner et al., 1983	NR	NR	NR	Ion fluctuation with naturally occuring weather conditions
Winsor and Beckett, 1958	NR	NR	NR	Ions generated by polonium foil with electrostatic field, 300 V potential used for ion separation

[a] NR = not reported.

Table 14
MICRO-CLIMATE ENVIRONMENTAL CHARACTERISTICS: HUMAN BIOMEDICAL RESPONSES

Study	T (°C)	RH (%)	BP (in.)	O_3 (ppm)	NO_x (ppm)	Air delivery system
Albrechtsen et al., 1979	NR[a]	NR	NR	NR	NR	NR
Dantzler et al., 1983	NR	NR	NR	NR	NR	NR
David et al., 1960	NR	NR	NR	NR	NR	NR
David et al., 1962	NR	NR	NR	Low	NR	Air-conditioned burn unit, ozone indicated to be not measureable above background but no quantitative data provided
Hawkins, 1981						
Site 1	24.3	48.4	NR	NR	NR	Sealed air centrally handled in modern office building, no information provided on filtration, air exchange, etc., O_3 levels not measured above ambient (levels not given)
Site 2	22.4	53.2				
Site 3	22.8	46.5				
Jones et al., 1976	NR	NR	NR	NR	NR	NR
Minehart et al., 1961	NR	NR	NR	NR	NR	NR
Motley and Yanda, 1966	21—24	NR	NR	NR	NR	Air filtered over activated carbon
Osterballe et al., 1979	NR	NR	NR	NR	NR	Air blown downward by ventilators at top of test chambers
Palti et al., 1966	NR	NR	NR	NR	NR	NR
Sulman et al., 1970	15—30	25	NR	NR	NR	Natural weather conditions known as Sharav in Israel
Sulman et al., 1971	15—30	25	NR	NR	NR	Same as Sulman et al., 1970
Sulman et al., 1973	15—30	25	NR	NR	NR	Same as Sulman et al., 1970
Sulman et al., 1974	28	25	NR	Low	Low	Reports of "minimum ozone and nitric oxides" and no "electrically charged aerosols"; no information provided regarding filtration, exchange, etc.
Sulman et al., 1975	NR	NR	NR	Low	Low	Same as Sulman et al., 1974
Sulman et al., 1978	18—30	20—70	NR	Low	Low	Ozone and nitric oxides measured and not detected above background; same as Sulman et al., 1974
Wagner et al., 1983	−1.6 to 9.4	NR	29.50—30.00	est.	NR	Natural changes in weather fronts, study period occurred with change from 20.6°C to −1.6°C along with precipitation
Winsor and Beckett, 1958	21—27	20—40	NR	NR	NR	Air velocity ranged from 10—30 ft/min

[a] NR = not reported; est. = estimated

Table 15
EXPERIMENTAL DESIGN, PROCEDURES AND STATISTICAL TESTS: HUMAN BIOMEDICAL RESPONSES

Study	Design	Grounding	Statistics	Replication	Other experimental issues/problems
Albrechtsen et al., 1979	Repeated measures	NR[a]	Friedman F test Wilcoxon matched pairs		Successive exposure to ions with no counterbalancing or detoxification periods resulting in potential confounding due to order or carryover effects
Dantzler et al., 1983	Repeated measures Double-blind	Yes	t tests X^2	No	Location of ion collector in the near vicinity of subject's head suggests likelihood that few if any ions were inhaled due to competition by collector
David et al., 1960	Case study	NR	NR	No	Subjects differed in degrees of burn, ambulatory condition, exposure period, and dosage level of ions, diet, medication, and experimental setting; no supportive data for conclusions reached, and no description of the comparable control group used
David et al., 1962	Independent two group	No	X^2	No	Possible placebo effects due to expectational set provided by experimenters (i.e., negative ions will improve breathing), statistics reported incorrectly for both tables, unclear criteria for patient classification regarding improvement, lack of air filtration information makes it impossible to determine potential physical environmental artifacts
Hawkins, 1981	Repeated measures	NR	F test X^2	No	Exposure period/week not reported (number of days or hours), insufficient explanation of data analysis procedures, effects of humidity and ions on headache confounded, no verification of uniform exposure for subjects within the test site
Jones et al., 1976	Repeated measures	NR	Mann-Whitney U F test X^2	No	Location of air vents and laminar flow may compete with ion transport to subject, raising exposure questions

Minehart et al., 1961	Indpendent two group	NR	X^2	No	Confounding factors present which were not statistically accounted for, (e.g., dependence of reduced pain reports on type of surgery, presence of suspicion and fear in some patients, and large difference in size of experimental and control groups); use of anecdotal subject reports of discomfort along with erratic exposure periods. Total exposure not reported
Motley and Yanda, 1966	Independent group Pre- and post-tests	Yes	NR	No	Data presented under subjective headings, (e.g., better/worse) are minimally reported, three different settings for different periods without any apparent rationale, no description given of subjects' age or sex
Osterballe et al., 1979	Repeated measures	NR	Nonparametric ANOVA Wilcoxon matched pairs	No	Possible carryover effects due to lack of intervals between exposures
Palti et al., 1966	Independent group Pre- and post-tests Factorial Repeated measures	NR	NR	No	Subjects were replaced periodically during the experiment so that only 4 out of 5 were identical for each treatment condition. Exposures repeated for some infants and not others, but all data were averaged together
Sulman et al., 1970	Epidemiological	No	NR	Partial	Distinction made between weather front (Sharav) and prefrontal period (1—2 days before Sharav), but data for individuals aggregated for both conditions, making interpretation difficult; only vague reference to actual number of Sharav days subjects were exposed to over 1-year period; clinical syndromes inadequately defined regarding number of symptoms experienced by each patient
Sulman, 1971	Clinical	No	NR	Partial	Same as for Sulman et al., 1970; highly subjective categories for data analysis, study is essentially pharmacological but no pre-post biochemical data given to illustrate the action of the drug administered to the subject
Sulman et al., 1973	Clinical	No	t test	Partial	Data given for 100 patients relieved by pharmacological intervention from Sharav-induced

Table 15 (continued)
EXPERIMENTAL DESIGN, PROCEDURES AND STATISTICAL TESTS: HUMAN BIOMEDICAL RESPONSES

Study	Design	Grounding	Statistics	Replication	Other experimental issues/problems
					symptoms of adrenal exhaustion, but no comparison data given for patients not relieved or for normal patients; differences in response based on age noted but no data provided
Sulman et al., 1974	Clinical Blind crossover	No	NR	Partial	Sample size for each experiment within the study unclear, breakdown showing relationship between subjects and symptoms missing, actual data showing 5-HT and 5-HIAA changes in response to ionization not reported
Sulman et al., 1975	Epidemiological Clinical Double-blind	No	t test	Partial	All-female subject population, and therefore difficult to compare with previous research, no breakdown showing percent of subjects experiencing a particular symptom within a given clinical syndrome
Sulman et al., 1978	Clinical	No	t test	Partial	Sample consisted of 5 female (weather-sensitive) and 5 normal males; rationale for this peculiar sample not given, making comparison with previous work difficult
Wagner et al., 1983	Field study	No	t test	Yes	Confounding with large changes in temperature, humidity and barometric pressure
Winsor and Beckett, 1958	Repeated measures	Yes	NR	No	Grounding variable and momentary

[a] NR = not reported.

studies, for example, were test subjects grounded, and, for one of the 2, grounding appears to be erratic.[75] None of the studies took any steps to ensure that clothing worn by test subjects had comparable electrostatic properties. As stated earlier, this is critical for controlling the ion exposure of individual test subjects. Moreover, only one researcher, Sulman, recognized the possibility that some subjects may be, for reasons not well understood, inherently sensitive to ions in varying degrees.

The usefulness of many studies (Table 11) is limited also due to a number of problems of execution and interpretation.[78,80-83] This is confounded further by a number of other elements of the Sulman studies, including minimal biochemical and environmental data. For example, the positive or negative ion concentration prevailing during pre-Sharav or Sharav periods was reported to be approximately 1200 ions/cm^3; by contrast, normal ambient ion concentrations on non-Sharav days were approximately 1000 ions/cm^3. This slight change in ion concentration is difficult to detect with accuracy as it is well within the likely range of measurement error using even state-of-the-art ion counters; and in any case could be considered an ambient level. Much more pronounced are the changes in temperature and humidity associated with the Sharav weather conditions, and it is likely the cluster of symptoms among weather-sensitive individuals reported by the Sulman group were produced by a complex interaction of these meteorological factors.

All of the clinical studies reviewed here contain numerous experimental artifacts and require replication under well-controlled conditions to determine their validity.

3. Results of Minimally Acceptable Studies

Of those studies examining effects of ions on clinical symptoms in humans (see Tables 11 through 15), three are of minimally acceptable scientific quality.[10,72,76] In the study by Hawkins,[10] office workers were exposed to negative ion concentrations (an average of 2.7 $\times$ 10^3/cm^3 for 4 to 6 weeks, 8 hr/day). Ion exposures were found to significantly reduce symptoms, although changes in environmental parameters other than ions produced effects as great or greater than the ions themselves. For example, Hawkins found that changes in temperature or relative humidity alone, with no ions present, resulted in effects greater than those due to ions. For example, ion exposures in the Hawkins study were found to significantly reduce symptoms — a $\geqslant$50% reduction in headache, and a $\geqslant$30% reduction in nausea and dizziness for some but not all of the subjects studied. By comparison, increases in mean temperature of 4°C reduced headache complaints by 51%, and a mean temperature increase of 9°C resulted in a 70% decrease in headache complaints. Similarly, nausea decreased by 90% with a 9°C rise in temperature, and dizziness decreased by 99% with the same rise in mean temperature. Furthermore, a decrease in relative humidity (average of 9%) was associated with a 14% decrease in headache, an 86% decrease in nausea, and a 100% decrease in dizziness. The second study meeting criteria of minimal acceptability reported improved lung function (as measured by forced expiratory volume FEV) with exposure to either positive (3.0 $\times$ 10^4/cm^3) or negative ions for 15 min.[72] The increases observed were extremely small, and were fully reversed 15 min after exposure. This study also reported no effects of either positive or negative ions on histamine-reduced broncho-constriction (or on subjective air quality rating).

In comparison to Sulman's previous studies, this later study used improved procedures, and the results are almost entirely negative.[76] In this study, no effects on various indicators of biomedical status were found in human subjects exposed for 2 months (6 hr/day) to negative ion densities (1.0 $\times$ 10^4 ions/cm^3). These indicators included blood pressure, pulse, respiration, glucose, and blood cell count. Overall, this study provides evidence for lack of biological effects of negative ions, excepting the data for effects on serotonin (50% reduction), although the generality of this study is limited because of the small sample size (ten subjects) used.

The weight of the results of the three studies meeting criteria for minimal scientific

Table 16
SUMMARY OF EXPERIMENTAL RESULTS: ANIMAL BIOMEDICAL RESPONSES

Study	Clinical symptoms		Air ion effects[a] (+)	Air ion effects[a] (−)
Deleanu et al., 1965	Ulceration		–	↓
Hinsull et al., 1981	Embryonic development			
	Differentiation		0	0
	Course of pregnancy		0	0
	Gestation time		0	0
	Neonatal development			
	Survival		0	↓
	Growth		0	0
	Chronic respiratory disease		0	↑
	Postweaning development		0	↑
Hinsull et al., 1983	Growth (body weight)		–	0
	Thymus weight		–	0↓
Hinsull et al., 1984	Growth (body weight)		–	0
Krueger and Levine, 1967	Mortality: Coccidioidomycosis		↑	–
Krueger et al., 1970	Mortality			
	Pneumonia		↑	–
	Influenza		↑	–
Krueger et al., 1971	Mortality: Influenza		–	0
Krueger and Reed, 1972	Mortality: Influenza	Low	↑	↑
		Medium	0	0
		High	–	↓
Krueger et al., 1974	Mortality: Influenza	Low	0	0
		High	0	0

[a] Key: – = not studied; 0 = no effects; ↑ = increase; ↓ = decrease.

acceptability indicates that exposure to positive or negative ions ranging in concentration from 10^3/cm^3 to 10^4/cm^3 has no effects on histamine threshold, body weight, blood pressure, pulse, respiratory rate, oral temperature, urinary pH, albumin, glucose, ketones, bilirubin, occult blood, blood cells, and ECG. Effects were observed on lung function, EEG, and 5-HT. In almost all cases these effects were small in magnitude.

D. Animal Biomedical Responses

Nine studies of the effects of air ions on animal pathology were reviewed. They examined the effects of exposure to air ions on ulceration, neonatal development, and respiratory disease in animals. The reported results indicate, though not categorically, that positive air ions exacerbated animal morbidity, while negative ions ameliorated it. The results of these studies are summarized in Table 16; the experimental characteristics and procedures of these studies are reported in Tables 17, 18, 19 and 20, respectively.

1. Reported Results: Aggregate Data

The most consistent result from this body of animal data, as stated above, is that positive ion exposure increased animal morbidity due to respiratory disease, while negative ion exposure decreased it. This pattern is best typified by the data from a series of five studies conducted from 1970 to 1974 by the Krueger group (see Table 16). Although these experiments are useful because they consistently used control groups to provide a relative index of the effects of air ion exposure, the data presented by the Krueger group are contradictory. For example, one study in the series found that ions did not alter the course of respiratory

Table 17
SUMMARY OF EXPERIMENTAL CHARACTERISTICS: ANIMAL BIOMEDICAL RESPONSES

Study	Air ion dosages (ions/cm³)		Exposure period (days)	
	(+)	(−)	(+)	(−)
Deleanu et al., 1965	−	$5.0 \times 15.0 \times 10^3$	−	(s) 3 (r) 15—24
Hinsull et al., 1981	1.0×10^4	1.0×10^4	14—15	14—45
Hinsull et al., 1983	−	1.0×10^4	−	28—140
Hinsull et al., 1984	−	1.0×10^4	−	35—140
Krueger and Levine, 1967	$3.0—4.0 \times 10^5$	−	30	−
Krueger et al., 1970	$1.0—2.0 \times 10^5$	−	6—10	−
Krueger et al., 1971	−	$2.0—5.0 \times 10^5$	−	11
Krueger and Reed, 1972				
Low	2.7×10^3	3.0×10^3	14—16	14—16
Medium	1.7×10^4	2.0×10^4	14—16	14—16
High	−	3.0×10^5	14—16	14—16
Krueger et al., 1974				
Low	5.0×10^3	5.0×10^3	11	11
High	5.0×10^5	5.0×10^5	11	11

disease (induced by delivering virus in aerosol form) at any ion concentration tested.[84] Krueger and Reed[85] on the other hand, reported that moderate dosages of positive and negative ions (on the order of 10^4 ions/cm³) did not affect mortality when virus was delivered by intranasal challenge, but that concentrations of 10^3 ions/cm³ or 10^5 ions/³ did; positive ions increased mortality due to respiratory disease, while negative ions decreased it. The nonlinearity of these data is confusing, making direct interpretation difficult.

Other data that concern the effects of ions on pathological conditions in animals are also problematic. One study reported that neonatal development was hampered significantly by exposure to negative ions, but unaffected by positive ions.[86] The same study also indicated that embryonic development was not altered by exposure to ions of either polarity. Subsequent attempts to replicate these data failed, with findings that negative ion exposure did not affect neonatal development (growth).[87,88]

Two different exposure periods, as noted in Table 17, were used in a study of the effects of negative ions on the development of ulceration in rabbits, one for (s) and one for (r).[89] These represent, respectively, starvation-induced (s) and reserpine-induced (r) ulceration. The differences in exposure periods (3 vs. 15 days) were due to the difference in the time necessary for ulceration to occur by these methods. In both cases negative ions were shown to reduce ulceration.

2. *Critique of Experimental Design and Documentation*

The studies with animals were reasonably well designed, but often poorly reported. As a result, patterns in the reported effects from ion exposure are discernible, but detailed interpretation of the results of any one study and reconciliation of apparent contradictions between various data are difficult at best.

The studies in this group made effective use of experimental control groups and displayed acceptable test environment control. Moreover, in eight of the nine studies, subjects were grounded, and in six of nine studies appropriate statistical analysis of results was undertaken (see Table 20).

None of the research with small animals adequately reported conditions of the test environment (see Tables 18 and 19). While nearly all of them indicated ion mobility, tem-

Table 18
DC ELECTRICAL ENVIRONMENTAL CHARACTERISTICS: ANIMAL BIOMEDICAL RESPONSES

Study	Field (V/cm)	Current density (A/cm²)	Mobility (cm²/V-s)	Ion exposure system
Deleanu et al., 1965	NR[a]	NR	1.0—2.0	Ions generated by alpha emitter and dispersed with airflow, ions generated by corona discharge in enclosed laminated case, ionization chambers constructed of "plywood laminated with formica and lined with aluminum" grounded to earth
Hinsull et al., 1981	NR	NR	NR	
Hinsull et al., 1983	NR	NR	0.05	Ions generated by corona discharge and measured with an ion counter
Hinsull et al., 1984	NR	NR	0.05	Same as Hinsull et al., 1983
Krueger and Levine, 1967	NR	NR	1.0—2.0	Ions generated by beta emitter (tritium absorbed on zirconium) with biased grid for polarity control, ions delivered to test animal by air and field, ions measured with ion counter, flow of current to ground (amp) was reported
Krueger et al., 1970	NR	NR	1.0—2.0	Same as Krueger and Levine, 1967
Krueger et al., 1971	NR	NR	1.0—2.0	Same as Krueger and Levine, 1967
Krueger and Reed, 1972	NR	NR	NR	Same as Krueger and Levine, 1967
Krueger et al., 1974				
High	40—60	NR	1.0—2.0	Same as Krueger and Levine, 1967
Low	1			

[a] NR = not reported.

Table 19
MICRO-CLIMATE ENVIRONMENTAL CHARACTERISTICS: ANIMAL BIOMEDICAL RESPONSES

Study	T (°C)	RH (%)	BP (in.)	O_3 (ppm)	NO_x (ppm)	Air delivery system
Deleanu et al., 1965	21—24	50—70	NR[a]	NR	NR	Air flow through chambers was 0.3 m/sec
Hinsull et al., 1981	21	50—60	NR	NR	NR	Negative pressure system
Hinsull et al., 1983	22 ± 1	60	NR	NR	NR	Air exchange 14 air charges/hr
Hinsull et al., 1984	22 ± 1	60	NR	NR	NR	Same as Hinsull et al., 1983
Krueger and Levine, 1967	NR	NR	NR	NR	NR	Pollution-free air, vacuum pan for waste collection
Krueger et al., 1970	19—25	48 ± 15	NR	NR	NR	Two different systems; first used double-thick acrylic fiber filter (ULOK, Union Carbide) with retention >10 μm; second used electrostatic, activated carbon, and absolute (atomic energy type) filters
Krueger et al., 1971	26 ± 1	45 ± 7	NR	NR	NR	Pollutant-free air delivered to test animals at 0.178 m^3/min; used electrostatic, activated carbon, and atomic energy-type absolute filters
Krueger and Reed, 1972	26 ± 1	45 ± 7	NR	est.	NR	Same as in Krueger et al., 1971
Krueger et al., 1974	26	40 ± 10	NR	est.	NR	Same as in Krueger et al., 1971

[a] Key: NR = not reported; est. = estimated.

perature, and relative humidity during testing, most failed to document electric field, current density, barometric pressure, and the presence of gaseous contaminants. The air filtration systems were described adequately in studies by the Krueger group, but not in the others (see Table 19).

Replication attempts among these studies where intranasal challenge was used showed some agreement between two separate studies that positive ions influenced the course of influenza in mice, resulting in greater mortality.[85,90,91] Effects of negative ions were less consistent. One study reported no differences in mortality between control animals and animals exposed to negative ions, while a second study showed that exposure to negative ions significantly decreased mortality. No explanation was provided by the investigators for this failure to replicate.

3. Results of Minimally Acceptable Studies

Of the studies reviewed concerning biomedical responses in animals, eight of the nine meet the criteria for minimally acceptable scientific quality.

In one study negative ions (1.5×10^4 ions/cm^3 for 3 to 24 days) significantly reduced ulceration in rats (induced either by starvation or reserpine injection) an average of 27%.[89]

A series of studies examining neonatal growth and development reported mixed results.[86-88] The experimental quality of the research by these investigators increased over time, with the experimentally soundest of these studies carried out in 1983 and 1984. In these latter studies, earlier reports of positive and negative ion effects on neonatal growth were not confirmed.

The other experimentally acceptable studies of ion effects of biomedical responses in animals were conducted by the Krueger group at Berkeley. These studies examined the effects of positive and negative ions on the rate of mortality in animals with respiratory disease. In the first of these studies, mortality was found to be significantly increased by exposure to positive ions (3.0 to 4.0×10^5 ions/cm^3) over 30 days.[92] Mortality in mice in a second study was similarly affected by positive ion exposure (1.0 to 2.0×10^5 ions/cm^3 for 6 to 10 days), although the difference between experimental and control animals was slight.[90] Using negative ion exposure (2.0 to 5.0×10^5 ions/cm^3 for 11 days), mice infected with the PR8 strain of influenza virus did not differ from controls in mortality rate.[91]

In a study examining several different concentration levels of negative or positive ions (low = 5.0×10^3 ions/cm^3, medium = 2.0×10^4 ions/cm^3 and high = 5.0×10^5 ions/cm^3), mice infected with a viral respiratory pathogen and exposed to ions for 14 to 16 days showed greater mortality relative to control in both negative and positive ion conditions, but only when the ion level was low. No effects on mortality were observed in animals exposed to medium levels of either polarity of ions. When a high negative ion concentration was used, mortality in mice was slightly lower than that of control animals. The effect of a high positive ion condition was not included.

In the last series of these studies, where viral respiratory pathogens were delivered in an aerosolized form, no effects of ions occurred on mice exposed to positive or negative ions at both high and low densities for 11 days.[84] However, the mortality among control animals used for this series of studies was significant, illustrating the high degree of variablity (error variance) within the experiment.

The data for effects of ions on respiratory disease show mixed results. The data do not follow a linear dose-related pattern, and as mentioned earlier, effects of ions appear to be weak.

In summary, the results of the research examining biomedical responses in animals are questionable due to the variability in the control group data, the uneven pattern of results, and the failure to replicate. If ion effects do exist, they are quite weak. In addition, full replication attempts examining effects of ions on neonatal development (see Table 16)[87,88] failed to support previously reported positive findings.[86]

Table 20
EXPERIMENTAL DESIGN, PROCEDURES AND STATISTICAL TESTS: ANIMAL BIOMEDICAL RESPONSES

Study	Design	Grounding	Statistics	Replication	Other experimental issues/problems
Deleanu et al., 1965	Factorial	NR	X^2	No	Ion mobilities reported as small and therefore likely to be estimated; no numbers given
Hinsull et al., 1981	Independent three group	Yes	NR[a]	No	Grounding of entire chamber yields a zero field at the intersection of the wall and floor yielding a zero ion concentration; uniformity of ion concentration is not specific; sensitivity units for ion measurement not explained; air filtration and airflow not specific
Hinsull et al., 1983	Independent two group	Yes	NR	No	Differences due to negative ion exposure reported but no statistical tests or p values reported
Hinsull et al., 1984	Independent two group	Yes	NR	No	Data reported contradicted Hinsull et al., 1981; attempt at full replication of previous study failed to reproduce earlier results
Krueger and Levine, 1967	Independent three group	Yes	X^2	No	
Krueger et al., 1970	Independent two group	Yes	X^2	Yes	Ion mobilities reported as small with no numbers given; two different series of studies carried out in two different physical environmental settings; replication of viral study in two different settings, but no replication of bacterial pneumonia series carried out
Krueger et al., 1971	Independent two group	Yes	X^2	No	Ion mobilities estimated; effects of negative ions not significantly different from control, although significantly different from positive ion effects reported in 1970
Krueger and Reed, 1972	Independent two group	Yes	X^2	Partial	Replication of effects of influenza virus using same strain as in Krueger et al., 1970; no explanation for data contradictory to those in Krueger et al., 1971, regarding no significant differences between high negative ions and control
Krueger et al., 1974	Factorial	Yes	X^2	No	Ion mobilities estimated, failure to replicate previous effects with same viral strain due to use of aerosol challenge in this study and intranasal challenge in previous studies

[a] NR = not reported.

III. CONCLUSIONS

The data reviewed in this chapter have examined behavioral and physiological effects of air ions using levels in the laboratory ranging from 10^3 ions/cm^3 (an essentially ambient level) to 10^7 ions/cm^3, with exposure periods ranging from minutes to months.

It appears that ion exposure is related primarily to behavioral and behaviorally relevant physiological changes in animals and humans. The kinds of behavioral and physiological effects most commonly observed in the literature are likely to be stimulus-dependent, meaning that the effects disappear when the stimulus is removed. To be emphasized is the fact that the recently reported data collected by the research group at The Rockefeller University consistently showed no effects of ions on either physiological (neurotransmitters in the brain) or behavioral responses.[33,34,62] These data are the result of a 3-year project using rigorous exposure conditions.

Throughout the review of the air ion literature, it was noted that much of it is not scientifically acceptable because poor experimental procedures were employed. Many studies did not report temperature, relative humidity, or electric field strength. Finally, the reported laboratory studies did not provide systematic dose-response and time-course data for small ion effects. Accordingly, this review identified those works which met certain minimum standards of scientific quality.

Based on a detailed examination of those studies meeting such minimum standards, the following conclusions seem reasonable: (1) there appear to be effects of ions on biological and behavioral responses in both animals and humans; (2) when these effects occur, they are for the most part small in absolute magnitude; and (3) the effects that have been observed are for the most part transient (the effects disappeared after subjects were removed from the ion exposure condition). It is clear from this review that in order for any future research to make a meaningful contribution to clarifying our understanding of biological effects of air ions, experiments must be carefully designed and rigorously executed, taking into account the exposure parameters identified in this chapter.

ACKNOWLEDGMENT

This chapter was supported in part by The Electric Power Research Institute under contract #RP1817-1.

REFERENCES

1. **Charry, J. M.,** Biological effects of small air ions: a review of findings and methods, *Environ. Res.*, 34, 351, 1984.
2. **Charry, J. M., Bailey, W. H., Shapiro, M. H., and Weiss, J. M.,** Ion exposure chambers for small animals, *Bioelectromagentics,* 7, 1, 1986.
3. **Weigel, R. J. and Kaune, W. T.,** Design, construction, and testing of a DC bioeffects enclosure for small animals, Final Report, Battelle Pacific Northwest Laboratories, Richland, Wash., 1985.
4. **Bracken, T. D.,** The HVDC transmission line environment, in Proc., Workshop on Electrical and Biological Effects Related to HVDC Transmission, Bracken, T. D., Ed., Battelle Pacific Northwest Laboratories, U.S. Department of Energy, Richland, Wash., 1979.
5. **Wehner, A. P.,** Special review electro-aerosols, air ions and physical medicine, *Am. J. Phys. Med.*, 48(3), 119, 1969.
6. **Albrechtsen, O., Clausen, V., Christensen, F. G., Jensen, J. G., and Moller, T.,** The influence of small atmospheric ions on human well being and mental performance, *Int. J. Biometeorol.*, 22(4), 249, 1978.

7. **Barron, C. I. and Dreher, J. J.,** Effects of electric fields and negative ion concentrations on test pilots, *Aerosp. Med.*, 35, 20.
8. **McGurk, F. C. J.,** Psychological effects of artifically produced air ions, *Am. J. Phys. Med.*, 38, 36, 1959.
9. **Hawkins, L. H. and Barker, T.,** Air ions and human performance, *Ergonomics*, 21(4), 273, 1978.
10. **Hawkins, L.,** The influence of air ions, temperature and humidity on subjective well being and comfort, *J. Environ. Psychol.*, 1, 279, 1981.
11. **Johnson, G. B.,** The electrical environment and HVDC transmission lines, in Proc. Conf. Environmental Ions and Related Biological Effects, Americna Institute of Medical Climatology, Philadelphia, 1982.
12. **Charry, J. M. and Hawkinshire, F. B. W.,** Effects of atmospheric electricity of some substrates of disordered social behavior, *J. Personality Soc. Psychol.*, 41, 185, 1981.
13. **Halcomb, C. G. and Kirk, R. E.,** Effects of air ionization upon the performance of a vigilance task, *J. Eng. Psychol.*, 4, 120, 1965.
14. **Ruocco, J. N.,** Effect of Ionized Air on Visual Reaction Time of Human Subjects, Ph.D. thesis, Fordham University, New York, 1962.
15. **Slote, L.,** An experimental evaluation of man's reaction to an ionized air environment, *Proc. Int. Conf. on the Ionization of the Air*, Vol. 2, American Institute of Medical Climatology, Philadelphia, 1961.
16. **Sovijarvi, A. R. A., Rosset, S., Hyvarinen, J., Franssila, A., Graeffe, G. and Lehtimaki, M.,** Effect of air ionization on heart rate and perceived exertion during a bicycle exercise test. *Eur. J. Appl. Physiol.*, 41, 285, 1979.
17. **Tom, G., Poole, M. F., Galla, J., and Berrier, J.,** The influence of negative air ions on human performance and mood, *Hum. Fact.*, 23, 633, 1981.
18. **Chiles, W. D., Cleveland, M. J., and Fox, R. E.,** A study of the effects of ionized air on behavior, WADD Technical Report No. 60-598, Nov., 1960.
19. **Hedge, A. and Eleftherakis, E.,** Air ionization: an evaluation of its physiological and psychological effects, *Ann. Occup. Hyg.*, 25(4), 409, 1982.
20. **De Vries, H. A. and Klafs, C. E.,** Ergogenic effects of breathing artifically ionized air, *J. Sports Med. Phys. Fitness*, 5, 7, 1965.
21. **Inbar, O., Rotstein, A., Dlin, R., Dotan, R., and Sulman, F. G.,** The effects of negative air ions on various physiological functions during work in a hot environment, *Int. J. Biometeorol.*, 26, 153, 1982.
22. **McDonald, R. D., Bachman, C. H., and Lorenz, C. J.,** Some psychomotor and physiological tests on humans exposed to air ions, *Aerosp. Med.*, 38, 145, 1967.
23. **Knoll, J., Rheinstein, J., Leonard, G. F., and Highberg, P. W.,** Influence of light atmospheric ions on human visual reaction time, *Proc. Int. Conf. on the Ionization of the Air*, Vol. 2, American Institute of Medical Climatology, Philadelphia, 1961.
24. **Koyranskiy, B. B. and Ukvol'berg, L. Y.,** The influence of air ionization on the ability to do mental work, *U.S. Jt. Publ. Res.* 10533, 1, 1961.
25. **Sigel, S.,** Bio-Psychological Influences of Air Ions in Men. Effects on 5HT and Mood, Ph.D. thesis, University of California, San Francisco, 1979.
26. **Charry, J. M.,** The Effects of Positive Air Ions of Human Performance, Physiology and Mood, Ph.D. thesis, New York University, 1976.
27. **Forward, J. R. and Alpern, H. P.,** Empirical evaluation of the effects of negative ion generators and electrostatic collectors on work-related stress behaviors, Study in the Social Services Building, Holland, Michigan, July 23—Sept. 9, 1980.
28. **Wofford, J. C.,** Negative ionization: an investigation of behavioral effects, *J. Exp. Psychol.*, 71, 608, 1966.
29. **Buckalew, L. W. and Rizzuto, A. P.,** Negative air ion effects on human performance and physiological condition, *Aviation, Space, and Environmental Medicine*, 1984, 731.
30. **Kelley, D. L.,** The Influence of Artificial Atmospheric Ionization on Human Motor Performance, Ph.D. thesis, University of Southern California, Los Angeles, 1963.
31. **Monaco, R. P. and Acker, C. W.,** Psychophysiological effects of ionized air on psychiatric patients, *Newsl. Res. Psychol.*, 5, 22, 1963.
32. **Yaglou, C. P., Brandt, A. D., and Benjamin, L. C.,** Observations on a group of subjects before, during and after expsoure to ionized air, *J. Ind. Hyg.*, 15(5), 341, 1933.
33. **Bailey, W. H., Charry, J. M., Weiss, J. M., Cardle, K., and Shapiro, M.,** Regional analysis of biogenic amine turnover in rat brain after exposure to electrically charged air molecules (air ions), *Soc. Neurosci.*, (abs.), 1983.
34. **Bailey, W. H. and Charry, J. M.,** Behavioral effects of air ions, U.S. DOE/EPRI Contractors Review Meeting, St. Louis, Mo., November 5—7, 1984(b).
35. **Frey, A. H.,** Modification of the conditioned emotional response by treatment with small negative air ions, *J. Comp. Physiol. Psychol.*, 63, 121, 1967.

36. **Nazzaro, J. R., Jackson, D. E., and Perkins, L. E.,** Effects of ionized air on stress behavior. *Med. Res. Eng.*, 50, 343, 1967.
37. **Falkenberg, V. and Kirk, R. E.,** Effects of ionized air on early acquisition of Sidman avoidance behavior by rats, *Psychol. Rep.*, 41, 1071, 1977.
38. **Lambert, J. F. and Olivereau, J. M.,** Single-Trial passive avoidance learning by rats treated with ionized air, *Psychol. Rep.*, 47, 1323, 1980.
39. **Olivereau, J. M. and Lambert, J. F.,** Effects of air ions on some aspects of learning and memory of rats and mice, *Int. J. Biometeorol.*, 25, 53, 1981.
40. **Bauer, F. J.,** Effects of ionized air and electroconvulsive shock on learning and innate behavior in rats, *Psychol. Monogr. Gen. Appl.*, 69, 1, 1959.
41. **Jordon, J. and Sokoloff, B.,** Air ionization, age and maze learning of rats, *J. Gerontol.*, 14, 344, 1959.
42. **Duffee, R. A. and Koontz, R. H.,** Behavioral effects of ionized air on rats, *Psychophysiology*, 1, 347, 1965.
43. **Terry, R. A., Harden, D. C., and Mayyasi, A. M.,** Effects of negative air ions, noise, sex, and age on maze learning in rats, *Int. J. Biometeorol.*, 13, 39, 1969.
44. **Diamond, M. C., Connor, J. R., Orenberg, E. K., Bissell, M., Yost, M., and Krueger, A.,** Environmental influences on serotonin and cyclic nucleotides in rat cerebral cortex, *Science*, 210, 652, 1980.
45. **Krueger, A. P. and Kotaka, S.,** The effects of air ions on brain levels of serotonin in mice, *Int. J. Biometeorol.*, 13, 25, 1969.
46. **Zemlan, F. P.,** Influence of p-chloramphetamine and p-chlorphenylalanine on female mating behavior, in Serotonin Neurotoxins, Jacoby, J. H. and Lytle, L. O., Eds., *Ann. N.Y. Acad. Sci.*, 305, 621, 1978.
47. **Kohler, C., Ross, S. B., and Srebro, B. and Ogren, S.,** Long-term biochemical and behavioral effects of p-chloramphetamine in rat, in Serotonin Neurotoxins, Jacoby, J. H. and Lytle, L. D., Eds., *Ann. N.Y. Acad. Sci.*, 305, 645, 1978.
48. **Browning, R. A., Hoffman, W. E., and Simonton, R. L.,** Changes in seizure susceptibility after intracerebral treatment with 5,7-dihydroxytyrptamine: role of serotonergic neurons, in Serotonin Neurotoxins, Jacoby, J. H. and Lytle, L. D., Eds., *Ann. N.Y. Acad. Sci.*, 305, 437, 1978.
49. **Hoebel, B. G., Zemlan, F. P., Trulson, M. E., MacKenzie, R. G., DuCret, R. P., and Norelli, C.,** Differential effects of p-chlorphenylalanine and 5,7-dihydroxytryptamine on feeding in rats, in Serotonin Neurotoxins, Jacoby, J. H. and Lytle, L. D., Eds., *Ann. N.Y. Acad. Sci.*, 305, 590, 1978.
50. **Messing, R. B., Pettibone, D. J., Kaufman, N., and Lytle, L. D.,** Behavioral effects of serotonin neurotoxins: an overview, in Serotonin Neurotoxins, Jacoby, J. H. and Lytle, L. D., Eds., *Ann. N.Y. Acad. Sci.*, 305, 480, 1978.
51. **Ellison, G. D.,** Animal models of psychopathology, *Am. Psychol.*, 32, 1036, 1977.
52. **Gilbert, G. O.,** Effect of negative air ions upon emotionality and brain serotonin levels in isolated rats, *Int. J. Biometeorol.*, 17, 267, 1973.
53. **Krueger, A. P., Andriese, P. C., and Kotaka, S.,** The biological mechanism of air ion action: the effect of carbon dioxide in inhaled air on the blood level of 5HT in mice, *Int. J. Biometeorol.*, 7, 3, 1963.
54. **Krueger, A. P., Andriese, P. C., and Kotaka, S.,** The effects of inhaling non-ionized or positively ionized air containing 2—4% carbon dioxide on the blood levels of 5HT in mice, *Int. J. Biometeorol.*, 10, 17, 1966.
55. **Krueger, A. P., Andriese, P. C., and Kotaka, S.,** Small air ions: their effect on blood levels of serotonin in terms of modern physical theory, *Int. J. Biometeorol.*, 12, 225, 1968.
56. **Olivereau, J. M.,** Effect of negative atmospheric ions on thermalgesia in mice, *Proc. Meet. Biology Society and its Affiliates*, 164, 287, 1970a.
57. **Olivereau, J. M.,** Behavior in mice subjected to thermo-algogenic stimulus after treatment with positive atmospheric ions, *C. R. Seances Soc. Biol. Ses Filiales*, 164, 501, 1970b.
58. **Lambert, J. F., Olivereau, J. M., and Tuong-Ngoc, A.,** Influence of artifical air ionization on the electroencephalogram of the awake rat, *Int. J. Biometeorol.*, 25, 71, 1981.
59. **Lacey, J. I.,** Somatic response patterning and stress: some revisions of activation theory, *Psychological Stress*, Appley, M. H. and Trumbull, R., Eds., Appleton-Century-Crofts, New York, 1967.
60. **Bachman, C. H., McDonald, R. D., and Lorenz, P. J.,** Some effects of air ions on the activity of rats, *Int. J. Biometeorol.*, 10, 39, 1966b.
61. **Olivereau, J. M.,** Influence of atmospheric ions on the activity of albino rats, *C. R. Seances Soc. Biol. Ses Filiales*, 164, 956, 1979.
62. **Bailey, W. H. and Charry, J. M.,** Measurement of neurotransmitter release and utilization in selected brain regions of rats exposed to DC electrical fields and atmospheric space charge, Interaction of Biological Systems with Static and ELF Electric and Magnetic Fields, 23rd Hanford Life Sci. Symp., Richland, Wash., October 2 to 4, 1984a.
63. **Bachman, C. H., McDonald, R. D., and Lorenz, P. J.,** Some physiological effects of measured air ions. *Int. J. Biometeorol.*, 9, 127, 1965.

64. **Bachman, C. H., McDonald, R. D., and Lorenz, P. J.,** Peak changes in electrocardiograms of rats exposed to ionized air, *Int. J. Biometeorol.*, 10, 101, 1966a.
65. **David, T. A., Minehart, J. R., and Kornblueh, I. H.,** Polarized air as an adjunct in the treatment of burns, *Am. J. Phys. Med.*, 39, 111, 1960.
66. **David, T. A., Dereau, J. V., Kornblueh, I. H., McGurk, F. C. J., and Minehart, J. R.,** Ionization of the air: the sedating effect of polarized air, in Biometeorology, Tromp, S. W., Ed., Pergamon Press, New York, 1962, 484.
67. **Minehart, J. R., David, T. A., McGurk, F. J., and Kornblueh, I. H., with Derau, J. V.,** The effect of artifically ionized air on postoperative discomfort, *Am. J. Phys. Med.*, 40, 56, 1961.
68. **Albrechtsen, O., Osterballe, O., and Weeke, B.,** Influence of small atmospheric ions on the airways in patients with bronchial asthma, *Symp. Indoor Climate*, 25, 377, 1979.
69. **Dantzler, B. S., Martin, B. G., and Nelson, H. S.,** The effects of positive and negative ions on bronchial asthma, *Ann. Allerg.*, 51, 362, 1983.
70. **Jones, D. P., O'Connor, S. A., Collins, J. V., and Watson, B. W.,** Effect of long-term air treatment on patients with bronchial asthma, *Thorax*, 31, 428, 1979.
71. **Motley, H. L. and Yanda, R.,** Environmental air pollution, emphysema and ionized air on psychiatric patients, *Dis. Chest*, 50, 343, 1966.
72. **Osterballe, O., Weeke, B., and Albrechtsen, O.,** Influence of small atmospheric ions on the airways in patients with bronchial asthma, *Allergy*, 34, 187, 1979.
73. **Palti, Y., DeNour, E., and Abrahamov, A.,** The effect of atmospheric ions on the respiratory system of infants, *Pediatrics*, 38, 405, 1966.
74. **Wagner, C. J., Danziger, R. E., and Nelson, H. S.,** Relation between positive small air ions, weather fronts and pulmonary function in patients with bronchial asthma, *Ann. Allergy*, 51, 430, 1983.
75. **Winsor, T. and Beckett, J. C.,** Biological effects of ionized air in man, *Am. J. Phys. Med.*, 37, 83, 1958.
76. **Sulman, F. G., Levy, D., Lunkan, L., Pfeifer, Y., and Tal, E.,** Absence of harmful effects of protracted negative air ionization, *Int. J. Biometeorol.*, 22(1), 53, 1978.
77. **Assael, M., Pfeifer, Y., and Sulman, F. G.,** Influence of artificial air ionization on the human electro-encephalogram, *Int. J. Biometeorol.*, 18, 306, 1974.
78. **Sulman, F. G., Levy, D., Pfeifer, Y., Superstine, E., and Tal, E.,** Effects of the Sharav and Bora on urinary neurohormone excretion in 500 weather-sensitive females, *Int. J. Biometeorol.*, 19, 202, 1975.
79. **Braker, W. and Mossman, A. L.,** *Gas Data*, Matheson, Lyndhurst, N.Y., 1980.
80. **Sulman, F. G., Danon, A., Pfeifer, Y., Tal, E., and Weller, C. P.,** Urinalysis of patients suffering from climatic heat stress (Sharav), *Int. J. Biometeorol.*, 14, 45, 1970.
81. **Sulman, F. G.,** Serotonin-migraine in climatic heat stress, its prophylaxis and treatment, *Proc. Int. Headache Symp.*, Elsinore, Denmark, 1971, 205.
82. **Sulman, F. G., Pfeifer, Y., and Superstine, E.,** Aldrenal medullary exhaustion from tropical winds and its management, *Isr. J. Med. Sci.*, 9, 1022, 1973.
83. **Sulman, F. G., Levy, D., Levy, A., Pfeifer, Y., Superstine, E., and Tal, E.,** Air-ionometry of hot, dry desert winds (Sharav) and treatment with air ions of weather sensitive patients, *Int. J. Biometeorol.*, 18, 313, 1974.
84. **Krueger, A. P., Reed, E. J., Day, M. B., and Brook, K. A.,** Further observations on the effect of air ions on influenza in the mouse. *Int. J. Biometeorol.*, 18(1), 46, 1974.
85. **Krueger, A. P. and Reed, E. J.,** Effect of the air ion environment on influenza in the mouse, *Int. J. Biometeorol.*, 16, 209, 1972.
86. **Hinsull, S. M., Bellamy, D., and Head, E. L.,** Effects of air ions on the neonatal growth of laboratory rats, *Int. J. Biometeorol.*, 25, 323, 1981.
87. **Hinsull, S. M., Head, E. L., and Bellamy, D.,** The effect of negative air ionization on the thymus glands of laboratory rats, *J. Biol. Phys.*, 11, 87, 1983.
88. **Hinsull, S. M., Bellamy, D., and Head, E. L.,** The effect of negative air ionization on the growth of four generations of laboratory rats, *Int. J. Biometeorol.*, 28, 163, 1984.
89. **Deleanu, M., Fritz, T., and Florea, E.,** Action of ionized air in the treatment of gastro-duodenal ulcers, *Int. J. Biometeorol.*, 9(2), 161, 1965.
90. **Krueger, A. P., Kotaka, S., Reed, E. J., and Turner, S.,** The effects of air ions on bacterial and viral pneumonia in mice, *Int. J. Biometeorol.*, 14(3), 257, 1970.
91. **Krueger, A. P., Kotaka, S., and Reed, E. J.,** The course of experimental influenza on mice maintained in high concentrations of small negative air ions, *Int. J. Biometeorol.*, 15, 5, 1971.
92. **Krueger, A. P. and Levine, H. B.,** The effect of unipolar positively ionized iar on the course of coccidioidomycosis in mice, *Int. J. Biometeorol.*, 11, 279, 1967.

Chapter 7

BIOLOGICAL RESPONSES TO AIR IONS: IS THERE A ROLE FOR SEROTONIN?

William H. Bailey

TABLE OF CONTENTS

I. INTRODUCTION

For the past 85 years scientists and other investigators have been searching for biological responses that could be produced by alterations in the electrical characteristics of the air. In this search, most interest has focused upon ionized air molecules (air ions). Hundreds of papers have been published concerning the possible influence of air ions on biological processes. Some weak influence of air ions has been reported in many of these studies and yet in a large number of other studies no influence whatsoever could be detected. Since no common thread has been identified to conceptually relate the apparently random findings in this literature, it is reasonable to doubt the hypothesis that air ions are biologically active.

Nevertheless, over the last 30 years, one area of air ion research, involving effects on serotonin [5HT] metabolism, has provided most of the data which has sustained the credibility of this hypothesis and continued to generate considerable interest and speculation.

The purpose of this chapter is to critically review the data and theoretical support for the hypothesis that air ions influence biological processes in animals by altering the release and metabolism of 5HT. The scope of this review is limited to the responses of laboratory animals to air ions. As yet, exposures to air ions and alterations in 5HT metabolism have not been convincingly associated in humans. The interested reader should therefore consult other recent reviews of these studies.[1,2]

II. LOCALIZATION AND FUNCTION OF 5HT

More than 10,000 papers have been published about 5HT since 1940 and so a great deal is known about the biochemistry and function of 5HT in biological processes. A brief overview of this area is included to give the reader a perspective with which to view the studies involving air ions that are reviewed later.

Serotonin was first extracted from the gastrointestinal tract in 1940.[3] Most of the body stores of 5HT are localized in the enterochromaffin cells of the intestinal mucosa, but it is found in a variety of other tissues in lesser amounts. Most of the 5HT appearing in blood is released from the gut, and vagal stimulation and increased intraluminal pressure can initiate its release. For instance, eating decreases the 5HT concentration in the stomach and duodenum of the rat by 30 to 40% and in humans elevates the concentration of 5HT in blood.[4,5] Conversely, fasting elevates the 5HT content of gastrointestinal tissue and reduces the concentration of 5HT in blood. 5HT that is released from tissues is rapidly metabolized to 5-hydroxyindole acetic acid (5HIAA) through the action of the enzymes, monoamine oxidase (MAO) and aldehyde dehydrogenase. The concentration of 5HIAA in urine or tissue is often used as an indicator of the activity of serotonergic systems.

Almost all of the 5HT not already metabolized to 5HIAA that is released into the bloodstream is sequestered in plateletes by an active uptake process. The 5HT that is taken up is stored in dense granule vesicles. Release of 5HT from these granules occurs during the aggregation of platelets in the early stages of blood clotting, and perfusion through the lung. As blood circulates through the lung, 5HT is removed from platelets. In fact, as much as 43% of the 5HT in platelets may be transferred to vascular endothelial cells during a single passage.[6] This uptake of 5HT by the lung has been hypothesized to play a role in modulating various aspects of pulmonary function that are regulated by smooth muscle.

Small amounts of 5HT in relatively high concentrations are found in the brain. Most of the serotonin in the brain is associated with the raphe nuclei in the upper pons and lower brainstem. Nerve fibers from these nuclei project widely throughout the brain and down the spinal cord. Depolarization of the nerve terminals on these cells leads to the release of 5HT. The diffusion of 5HT across synaptic junctions in turn affects the electrical potential and function of adjacent neurons by binding to specialized receptors.

The highest concentration of 5HT observed anywhere in the body is in the pineal gland. This small gland, although attached to the brain, is innervated by sympathetic nerve fibers originating in the superior cervical ganglion in the neck. The pineal, unlike other tissues in the body, metabolizes 5HT to melatonin rather than to 5HIAA. The production of melatonin is largely controlled by the activity of *n*-acetyltransferase (NAT). In the rat pineal the level of 5HT is high during daylight and NAT activity is low, whereas in darkness NAT activity increases dramatically, increasing the availability of *n*-acetylserotonin for the synthesis of melatonin by hydroxyindole-*o*-methyltransferase (HIOMT). Human pineal function also exhibits a diurnal rhythm in the secretion of melatonin. The pineal gland, and particularly the melatonin it secretes, have been suspected as having effects on diverse biological systems, but their influence on hypothalamic-pituitary system control over endocrine function has been most intensively investigated.

III. ORIGINS OF THE 5HT HYPOTHESIS-MUCOCILIARY ACTIVITY IN THE TRACHEA

Krueger and his colleagues conducted a series of studies on the response to air ions of isolated tracheal tissue from five species.[8-12] They reported that positive air ions: (1) reduce "the frequency of ciliary movement in the trachea and mucous secretion; (2) contract and vasoconstrict the tracheal wall; (3) exaggerate the mucosal response to injury; and (4) increase respiratory rate." The effects of positive air ions on rabbit and guinea pig tracheal tissue were similar to those produced by the intravenous administration of 5HT to the animal. The administration of iproniazid, a drug that increases tissue levels of 5HT by inhibiting its degradation of 5HIAA via MAO was also reported to mimic the responses observed following exposure to positive air ions.[10] The interpretation of this latter finding is unclear for iproniazid is not specific in its action; it not only increases tissue levels of 5HT, but also increases tissue levels of norepinephrine, dopamine, and other biologically active compounds that are metabolized by MAO.

In contrast to the effects of positive ions on tracheal function, Krueger and his colleagues reported that negative ions increased ciliary activity, mucous flow, and reduced respiration, while normalizing the vasoconstriction and sensitivity of tracheal tissue to injury that occurred with exposure to positive ions. A decline in the tissue content of 5HT was also observed to occur for rabbit trachea exposed in vitro or in vivo to negative ions.[11] This depletion of 5HT was thought to be a critical factor in explaining the effect of negative ions on tracheal function because of the similar response of the trachea to reserpine. However, reserpine, like iproniazid, does not selectively affect cells containing 5HT, and it produces large depletions of norepinephrine and dopamine as well. Reserpine depletes tissues of 5HT and other neurotransmitters by inhibiting neurotransmitter binding within synaptic vesicles. This leads to 'leakage' of neurotransmitter from vesicles and rapid degradation.

Perhaps the most surprising aspect of this work, however, was the finding that ciliary reponses to negative ions were reported to persist for a month even though an increased excretion of 5HIAA in urine was detectable only during the first of several 24-hr exposures spaced over a period of several days.[11] In other studies it was claimed that 24 hr of exposure to air ions was not necessary to produce a long-term effect and that even single 20 to 60 min exposures to negative or positive air ions altered the ciliary rate for 1 month.[12]

The above data were interpreted by Krueger as indicating that high concentrations of air ions reliably influence mammalian mucociliary function through an effect on 5HT metabolism. Subsequent investigations in three different laboratories, nevertheless, failed to replicate any of the reported effects of air ions on ciliary rate and mucous secretion described by Krueger.[13-16] Krueger's findings appear artifactual and attributable to inappropriate experimental technique. Andersen has explained in detail that Krueger did not closely control temperature, humidity, or air quality in his experiments.[15] Control over these factors, as described by Andersen[16] in his own elegant treatise on this problem, is absolutely essential because tracheal tissue is exquisitely sensitive to even very small perturbations of these parameters. Therefore, air ions appear not to have been solely responsible for the effects observed on tracheal tissue, a conclusion with which Krueger himself has reluctantly concurred.[17] The validity of Krueger's reports is further called into question with regard to his finding that 5HT inhibited tracheal function, for again other investigators have reported that 5HT has either no effect, or a stimulatory effect on ciliary activity.[15,18]

Thus, the data which originally suggested that positive and negative air ions exert opposite effects on mammalian tissue through their influence on 5HT metabolism cannot be considered as evidence for any specific effect on 5HT metabolism or other biological effects of these electrically charged air molecules. It also should be noted that in all but one of Krueger's studies of air ion exposures on the trachea,[9] the ion exposures to the trachea in vitro were very much greater than could occur with exposures in vivo.

IV. AIR ION EXPOSURES AND 5HT METABOLISM

A. 5HT in Blood

Based upon his earlier studies on the trachea, Krueger expanded his work to determine if air ions affected 5HT metabolism in other tissues. He reported that mice exposed to positive ions constituted from purified air (4.3×10^5 ions/cm^3) increased blood 5HT levels by 24% above pre-exposure levels after 7 days.[19] Positive ionization of CO_2 alone when added to the environment also increased blood 5HT levels after 3 days (39%) and 7 days (22%) of exposure. Control animals exposed to untreated air showed increased blood 5HT levels at these same exposure periods of 21 and 13.5%, respectively. Of the mice exposed to high concentrations of CO_2+, 20% became ill and died. Small hemorrhages in the lungs and intestine were observed at autopsy. Exposure to high concentrations of negatively charged O_2, uncharged CO_2, or lower concentrations of CO_2+ (2.3×10^3 ions/cm^3) did not affect blood 5HT levels.

In a later experiment Krueger reported that increasing the concentration of CO_2 in the exposure chamber to 2 or 4% CO_2 without ionization reduced blood 5HT levels after 3 to 12 days.[20] Concomitant positive ionization of the chamber (4.3×10^5 ions/cm^3) prevented the fall in blood 5HT and increased the 5HT levels above pre-exposure levels after 14 and 22 days of exposure. A higher concentration of positive ions (5×10^5 ions/cm^3) increased blood 5HT levels in mice exposed to 4% CO_2 within as few as 3 days. No illness was reported in this study for mice exposed to positively ionized air containing high concentrations of CO_2 (2 to 4%) so that the illness and mortality observed by Krueger in his earlier experiment[19] probably occurred when essentially all the gas molecules were CO_2+. At ambient levels (0.05%) or even toxic concentrations (4%) of CO_2 in the atmosphere, the concentrations of CO_2+ molecules in Krueger's experiment were apparently too small to produce adverse effects. Positive ionization of O_2 and N_2 gases prior to addition to the exposure chamber was also reported to increase blood 5HT levels in mice as was observed with CO_2 but without ill effects.[20]

Negative air ions were reported to produce responses opposite to those of positive ions. Negative ions at concentrations of 4 to 5×10^5 ions/cm^3 decreased the concentration of 5HT in mouse blood within 1 to 3 days, and lowered levels of blood 5HT persisted up to 7 days in mice exposed to negatively charged O_2 or CO_2 at similar concentrations.[21]

There is less concern than for most of the other studies reported in the air ion literature that the animals in these studies were, in fact, exposed to air ions because of attention to, and control over, some important technical parameters. Nevertheless, the data from these experiments are open to question on other grounds. The blood 5HT values were extremely variable and relatively large numbers of mice (20 to 80) were tested to produce small enough standard errors to obtain statistically significant effects which were as a rule quite modest (6 to 18% differences between experimental and control mice). Also, the chemical method used by Krueger to measure 5HT (the Udenfriend method) is generally recognized as neither particularly specific nor sensitive.[22] There is also a question as to whether the alterations in the concentration of 5HT in blood Krueger reported were associated with an increase in the concentration of 5HT in platelets or in the number of platelets/mℓ of blood. Finally, variations in blood levels of 5HT within the ranges observed by Krueger occur naturally as a function of diet, gut motility, etc., so that it is possible that the variations in blood 5HT he observed were only a marker (i.e., an indirect effect) of the influence of air ions on some other biological process.

We have investigated these findings in our laboratory under well-characterized and validated exposure conditions. We exposed rats to positive or negative ions (14 to 18 per group) at a concentration of 5.0×10^5 ions/cm^3 for 18 hr.[23] Particular care was taken to prevent exposure of the rats to corona discharge products — ozone, oxides of nitrogen, and noise — whose effects might be mistakenly attributed to air ions.[24] Platelets were isolated from blood and counted. The concentration of 5HT in platelets was measured by high pressure liquid chromatography. No differences in platelet count or in 5HT content were found between animals exposed to air ions and unexposed controls.

Longer-term studies recently have been reported in which rats were exposed to negative or positive ions at 1.5×10^6 ions/cm^3 for 21 days. No significant changes in the level of (free or total) tryptophan, the amino acid precursor to 5HT, or in the level of 5HT in blood were observed.[25] Even longer — lifetime — exposures to 2×10^3 ions/cm^3 or 2×10^5 ions/cm^3 of either polarity have not been found to affect any measurements of blood 5HT that were taken at regular 3-month intervals (eight samplings) during a 2-year study.[26,27]

Although a wide range of air ion exposure levels and exposure durations have been investigated to establish a relationship between air ions and blood 5HT, all studies that have been published since Krueger's initial studies (1963 to 1968) have reported negative results. This strongly argues against the possibility that air ions exert any significant direct or indirect biological effects that are reflected in altered blood 5HT levels.

B. 5HT in Brain

Keueger extended his studies of the effects of air ions on 5HT metabolism to brain tissue.[28] Mice were exposed to low (2 to 4×10^3 ions/cm^3), medium (3 to 4×10^4 ions/cm^3), and high concentrations (3.5 to 5×10^5 ions/cm^3) or positive or negative ions for 12, 24, 48, and 72 hr. At the lowest ion concentration, only positive ions at 12 hr and both positive and negative ions at 72 hr affected brain 5HT content (a decrease of 9 to 11%). In mice exposed to medium concentrations of air ions, only positively charged ions at 12 hr and negatively charged ions at 72 hr reduced brain 5HT levels (7 to 8%). At the highest ion concentration, both positive and negative ions reduced brain 5HT levels after 12 hr and 72 hr of exposure (10 to 11%). These reductions in brain 5HT did not appear to be consistently related to the polarity of charge on the ion, or the ion concentration. However, both positive and negative ions were more likely to decrease brain 5HT after the longest exposure period, 72 hr.

A study by Gilbert has been often quoted as confirming Krueger's finding that air ions reduce brain 5HT levels.[29] Gilbert exposed rats housed in individual cages and group cages to negative ions (3×10^3 ions/cm^3) continuously or intermittently for 100 days. Ion exposure did not affect body weight, fear avoidance responses, or runway activity. Intermittent ion exposure of group-housed rats or individually housed rats had no effect on the concentration of 5HT in brain. The concentration of 5HT was represented as being significantly decreased in individually housed rats that were continuously exposed to negative air ions. However, when one checks the t value and evaluates it in the appropriate nondirectional (two-tail) t-test, the effect of ionization is not statistically significant in this condition.

A much greater effect of negative air ions on brain 5HT levels was later demonstrated by Diamond, Krueger, and their colleagues.[30] Rats exposed to 1×10^5 ions/cm^3 in a multifamily enriched environment for 20 days beginning at age 6 days were reported to have lower concentrations of 5HT (45 to 61%) and adenosine 3′,5′-monophosphate (cAMP) (35 to 45%) in somatosensory and occipital cortex than unexposed controls. In contrast, the brain weights of rats exposed to negative ions were 11 to 12% greater than those of controls. The concentration of 5HT in the cortex of rats reared in a single-family environment within ordinary laboratory cages was not affected by exposure to negative ions although the cortical weights of ion-exposed rats in this condition were greater (17%) and the cAMP concentrations were lower (30%) than those of controls.

The effect of negative ions on brain 5HT levels demonstrated in this study for rats under some housing conditions was certainly larger and more convincing (appropriate statistical evaluations were made and a more specific assay method for 5HT was employed) than was reported in the earlier Krueger experiments. Nevertheless, the manner in which the brain samples were obtained for the assay of 5HT and cAMP clouds the interpretation of these findings. Alterations in the cAMP content of the brain can occur within a few seconds and so, to prevent this, measurements should always be made on brain tissue that has been rapidly frozen in liquid nitrogen or subject to enzyme inactivation by focused microwave irradiation to the skull before the brain is removed. However, in this study a delay of 4 min was reported to have occurred before the dissected brain tissue was frozen. This delay certainly altered the levels of cAMP and very likely altered the levels of 5HT that were reported, and it cannot be ruled out that the reported changes in these parameters occurred postmortem.

Other aspects of this experiment also not discussed by the authors are worth noting. As described above, in the multifamily enriched environment the concentration of 5HT in the cortex of rats exposed to negative ions was 45 to 61% lower than was observed in rats exposed to atmospheric air. However, the magnitude of this difference was comparable to the 60% difference in the concentration of 5HT observed in visual cortex between rats raised in multifamily environments and rats raised in single-family environments, both of whom were only exposed to atmospheric air. Differences between these two housing conditions also produced a 34% difference in the levels of cAMP found in the somatosensory cortex. These examples illustrate the point that large alterations in brain 5HT or other neurotransmitter levels like those reported by Krueger and Diamond can be produced by simple alterations in an animal's environment, not just air ions. Another point is that dramatic changes in the concentration of 5HT in brain are, like the concentration of 5HT in blood, every day occurrences in the life of a rat, mouse, or for that matter, man. The concentration of 5HT in brain in rats and other species is subject to a circadian rhythm resulting in large variations during the light-dark cycle. For example, the concentration of 5HT in rat brain has been reported to decline by as much as 56% during the first 2 hr of darkness in some brain regions and then increase over the following 18 to 24 hr.[31] Thus, if exposure to air ions were to produce a phase shift in the diurnal rhythm of the animal (possibly for reasons unrelated to any specific effect on 5HT), then comparisons with control animals could yield significant differences in 5HT levels in brain and blood for this reason alone.

In our laboratory we also have investigated the effects of exposure to air ions on serotonin metabolism in brain.[32,33] Rats were exposed individually to 5.0×10^5 ions/cm^3 of either positive or negative charge, to dc electric fields of the same magnitude as that measured in the ion-exposure conditions (3.0 kV/m), or to ambient control-conditions for 2, 18, or 66 hr. The exposure system was the same as that used in our platelet study described previously.[24] At the end of each exposure period the rats were removed form the chambers and the concentration of 5HT and 5HIAA were measured in five brain regions. No effects of air ion or electric field exposure were found on 5HT levels, 5HIAA levels, or on the ratio, 5HIAA/5HT — a measure of neurotransmitter turnover, in any of the brain regions even when the p value for statistical significance was relaxed to $p < 0.10$.

Another recent study of rats exposed to air ion has suggested apparently quite different results. Beardwood reported that 5HT levels in brain were elevated (83%) by exposure to positive ions and 5HT levels were decreased by exposure to negative ions when 7-day exposures at ion concentrations of 7×10^5 were employed.[34] Air ion exposures also were reported to produce alterations in pain-mediated spinal reflexes thought to be mediated by descending 5HT pathways. These findings should be interpreted cautiously. The brain samples were apparently removed for the analysis of 5HT after repeated testing for responses to painful stimulation, and the method used to quantitate 5HT was the same nonspecific Udenfriend method used by Krueger to measure 5HT in his blood and brain samples. No details concerning the exposure system have been described.

Although the 7-day exposures employed by Beardwood were longer than those of other studies reporting variable, or no effects, exposures of longer duration do not seem to increase the likelihood that air ions will affect brain levels of 5HT. Dowdall and de Montigny[25] exposed rats to positively or negatively charged air ions (1.5×10^6 ions/cm^3) for 21 days without finding any effect of exposure on 5HT, or the metabolically related compounds, tryptophan (the amino acid precursor to 5HT) and 5HIAA. Since the brain samples were apparently taken from anesthetized rats at the conclusion of electrophysiological studies (see below), the generality of these findings, however, may be limited.

Dowdall and de Montigny did report a more interesting finding in this study — that exposure to air ions significantly affected a circadian rhythm in the sensitivity of hippocampal brain cells to iontophoretically applied 5HT. During the winter there is a circadian rhythm in the sensitivity of cells from control animals to 5HT; sensitivity increases from morning to evening. In the summer, however, this variation in sensitivity does not occur. Exposure to negative ions was observed to increase the sensitivity of neurons to 5HT in both winter and summer. In the winter sensitivity to 5HT increased particularly during morning hours, and so no circadian variation in neuronal sensitivity was observed in rats exposed to negative ions in this season. An opposite response, a decrease in the sensitivity of neurons to 5HT, occurred with exposure to positive ions, especially in the summer. No circadian variations in sensitivity to 5HT were observed in these animals or in the negative-ion and control groups during the summer. A circadian variation was observed in rats exposed to positive ions during the winter, but it was opposite to that noted in control rats; responses to 5HT steadily increased rather than decreased from morning to evening. No circadian rhythm was observed in rats exposed to negative air ions in the winter. Responses of hippocampal neurons to norepinephrine and acetylcholine did not vary during the day, nor were responses to these neurotransmitters affected by exposure to positive or negative air ions.

In assessing these results, attention should be given to the exposure system. Few details of the experiment were described or measured. However, there is information that the rats were placed in the exposure cage only 30 cm below the ionizing electrode. It is therefore possible that while attempting to climb out of the cage that the rats could be shocked by touching the electrode. However, this aspect of the experiment, like other possible confounding factors such as ozone and corona noise that may be present in many air ion experiments, cannot easily explain how negative and positive ions differentially affected the circadian-determined response of hippocampal neurons to 5HT.

Assuming that the findings are not otherwise artifactual, there is another plausible explanation. As was pointed out before, the rats are in close proximity to the ionizing electrode and so they are exposed also to light and other products of corona activity that are generated, not just air ions. Since the responses of hippocampal neurons to 5HT were demonstrated to be exquisitely sensitive to diurnal and seasonal lighting cues, perhaps the exposure to corona-generated light selectively altered the circadian rhythm underlying the responsiveness of neurons to 5HT. Air ions might also be capable of influencing these circadian rhythms. However, in view of the lack of any demonstrated mechanism by which air ions could produce such an effect, light appears to be a more likely agent in this experiment. Differential responses to positive and negative air ions might relate to differences in the wavelength or intensity spectrum of light generated at positively and negatively charged electrodes. Because of the circadian rhythm in the demonstration of the reported air ion effects, the possible involvement of the pineal gland should be considered. The pineal plays an important role in modulating circadian rhythms in the body, particularly those involving neural and endocrine systems. An involvement of the pineal in responses to air ions has not been previously investigated. However, the pineal appears to be affected by other electrical stimuli. For instance, 30-day exposures to 1.7 to 1.9 kV/m ac electric fields reduce circadian variations in melatonin biosynthesis in the rat pineal such that the increase in melatonin levels typically observed during darkness do not occur.[35,36]

C. Exposure and Dosimetry Considerations

Air ions have not been shown to reliably influence any biological parameter as yet and so there is no empirical basis to determine a mechanism by which exposure to air ions would lead to biological responses. However, given the physical characteristics of air ions, interactions of air ions with organisms appear only feasible at the air-body surface interface. Furthermore, it seems unlikely that air ions would significantly impact the external body surface sufficiently to influence the internal physiologic milieu except in trivial ways, fur or hair charging, for example. Thus, by exclusion, one is left with the possibility that responses to air ions are mediated at an internal body surface, e.g., oral, nasal, or respiratory membranes.

Indeed, there is some weak experimental evidence to support the hypothesis that the ions must be accessible to nasal or oral respiratory passages to affect animals or humans. For example, Bachman and colleagues[37] have reported that stimulatory effects of exposure to positive and negative ions on heart rate and respiration in rats were not observed, or even reversed, when ions were presented only to the outer surface of the nose, but could not be inhaled. Also, Knoll et al.[38] indicated that the very small effects of air ions on visual reaction time that they reported were demonstrable only if the human subject breathed through his mouth.

Realistic limitations as to how air ions could interact with organisms as outlined above, in turn, place limitations on the ways in which air ions could possibly affect serotonin metabolism. In consideration of the extremely low concentrations of air ions (10^5 ions/cm^3 yields a concentration of one air ion in 10^{14} air molecules), it seems most likely that air ions have to act at specialized sites on mucous membranes, i.e., at receptors, if there were any chance for biological effects to occur at such extremely low concentrations. One possibility is that air ions stimulate the chemoreceptors in nasopharangeal and pulmonary tissue which influence respiration rate, blood gas composition, and pH. Activation of vagal nerves by such stimuli could lead to the release of 5HT from enterochromaffin cells in the gut. The possibility was discussed earlier in this chapter that air ions, or light from corona activity, might alter circadian rhythms in the pineal gland. These stimuli might be transmitted to the pineal via vagal (for air ions) or visual (for light) pathways. Air ions also might directly influence 5HT metabolism in the lung and in this way possibly exert an influence over smooth muscle control of airway resistance, for 5HT is known for its effects on smooth muscle elesewhere. A hypothesis as to how air ions could affect serotonin metabolism via receptors in the respiratory system is discussed further in Chapter 8 by Robert Kavet.

V. CONCLUSION

The hypothesis that air ions influence 5HT metabolism was originally suggested on the basis of studies of mucociliary function. However, the reported effects of air ions on mucociliary function which prompted this hypothesis appear to be spurious. Rats and mice have been examined following exposure to air ions over a wide range of concentrations and exposure periods for evidence that 5HT levels in blood are affected. The bulk of the evidence does not support such an association. Although the results of some studies suggest that 5HT levels in brain may be reduced following exposure to air ions, the failure to obtain a similar effect in carefully controlled exposure conditions, and the presence of various confounding variables, cast doubt upon the certainty of such findings. Exposure and dosimetry considerations also severely limit the possible mechanisms by which air ions might influence 5HT metabolism. A recent study has suggested that exposure of rats to air ions generated by corona discharge modifies a circadian rhythm in the responsiveness of brain neurons to 5HT. The possibility is discussed that this effect might be mediated by light produced by corona rather than air ions. Such questions will continue to plague this and other areas of air ion research, until careful attention to confounding factors, through characterizations of the exposure conditions, and adequate control groups are recognized to be of the utmost importance. Without these ingredients, little confidence can be placed in reported associations of exposure to air ions and biological responses.

ACKNOWLEDGMENTS

The research conducted in our laboratory has been supported by the Electric Power Research Institute (Contract RP 1817). Special appreciation is extended to Drs. Robert Kavet and Robert Patterson for their assistance.

REFERENCES

1. **Bailey, W. H., Bissell, M., Brambl, R. M., Dorn, C. R., Hoppel, W. A., Sheppard, A. R., and Stebbings, J. H.,** *A Health & Safety Evaluation of the ± 400 kV Powerline,* Minnesota Environmental Quality Board, 1982, chap. 3.
2. **Bailey, W. H.,** Biological effects of air ions on serotonin metabolism: Fact and Fancy, in *Conf. Environ. Ions and Related Biological Effects,* Charry, J. M., Ed., American Institute of Medical Climatology, Philadelphia, 1982.
3. **Ersparmer, V.,** Pharmakologische Studien über Enteramin, *N-S Arch. Exp. Pathol. Pharmakol.,* 196, 343, 1940.
4. **Biggio, G., Piccard, M. P., Porceddu, M. L., and Gessa, G. L.,** Changes in gastro-intestinal serotonin content associated with fasting and satiation, *Experientia,* 33, 745, 1977.
5. **Warner, R. R. P.,** Current status and implication of serotonin in clinical medicine, in *Advances in Internal Medicine,* Dock, W. and Snapper, I., Eds., Yearbook Medical Publishers, Chicago, 1967.
6. **Wiersma, D. A. and Roth, R. A.,** Clearance of 5-hydroxytryptamine by rat lung and liver: the importance of relative perfusion, *J. Pharm. Exp. Ther.,* 212, 97, 1980.
7. **Krueger, A. P. and Smith, R. F.,** Effects of air ions on isolated rabbit trachea, *Proc. Soc. Exp. Biol. Med.,* 96, 807, 1957.
8. **Krueger, A. P. and Smith, R. F.,** The effects of air ions on the living mammalian trachea, *J. Gen. Physiol.,* 42, 69, 1958.
9. **Krueger, A. P., Smith, R. F., and Millar, J.,** Effects of air ions on trachea of primates, *Proc. Soc. Exp. Biol. Med.,* 101, 506, 1959.
10. **Krueger, A. P. and Smith, R. F.,** The biological mechanism of air ion action. I. 5-Hydroxytryptamine as the endogenous mediator of positive air effects on the mammalian trachea, *J. Gen. Physiol.,* 43, 533, 1960.

11. **Krueger, A. P. and Smith, R. F.,** The biological mechanism of air ion action. II. Negative air ion effects on the concentration and metabolism of 5-hydroxytryptamine in the mammalian respiratory tract, *J. Gen. Physiol.,* 44, 269, 1960.
12. **Krueger, A. P. and Smith, R. F.,** Parameters of gaseous ion effects on the mammalian trachea, *J. Gen. Physiol.,* 42, 959, 1959.
13. **Kensler, C. J. and Battista, S. P.,** Chemical and physical factors affecting mammalian ciliary activity, *Am. Rev. Respir. Dis.,* 13, 93, 1966.
14. **Guillerm, R., Badre, R., Hee, J., and Razouls, C.,** Effects des ions lesers atmospheriques sur l'activite ciliaire de la mugheuse tracheale de moutar et de Lapin in vitro, *C. R. Acad. Sci. (Paris),* D262, 669, 1966.
15. **Andersen, I.,** Effects of natural and artificially generated air ions on mammals, *Int. J. Biometeorol.,* 5, 229, 1972.
16. **Andersen, I.,** *Mucocilary Function in Trachea Exposed to Ionized and Non-ionized Air,* Ph.D. thesis, Akademisk Boghandel, Aahrus, Denmark, 1971.
17. **Krueger, A. P.,** Are air ions biologically significant? A review of a controversial subject, *Int. J. Biometeorol.,* 16, 313, 1972.
18. **Tsuchiga, M. and Kensler, C. J.,** Effects of autonomic drugs and 2-amino-methyl propanol, choline antagonist on ciliary movement, *Fed. Proc.,* 18, 453, 1959.
19. **Krueger, A. P., Andriese, P. C., and Kotaka, S.,** The biological mechanism of air ion action: the effect of CO_2+ in inhaled air on the blood level of 5-hydroxy-tryptamine in mice, *Int. J. Biometerol.,* 7, 3, 1963.
20. **Krueger, A. P., Andriese, P. C., and Kotaka, S.,** The effects of inhaling non-ionized or positively ionized air containing 2—4% carbon dioxide on the blood levels of 5HT in mice, *Int. J. Biometeorol.,* 10, 17, 1966.
21. **Krueger, A. P., Andriese, P. C., and Kotaka, S.,** Small air ions: their effect on blood levels of serotonin in terms of modern physical theory, *Int. J. Biometeorol.,* 12, 225, 1968.
22. **Maickel, R. P.,** Analytical procedures for serotonin and related compounds in biological samples, in *Serotonin in Health and Disease,* Vol. 1, Essman, W. B., Ed., Spectrum Publications, New York, 1978.
23. **Bailey, W. H., and Charry, J. M.,** unpublished data, 1983.
24. **Charry, J. M., Shapiro, M. H., Bailey, W. H., and Weiss, J. M.,** Ion-exposure chambers for small animals, *Bioelectromagnetics,* 7, 1, 1985.
25. **Dowdall, M. and De Montigny, C.,** Effect of atmospheric ions on hippocampal pyramidal neuron responsiveness to serotonin, *Brain Res.,* 342, 103, 1985.
26. **Kellogg, E. W., Yost, M. G., Reed, E. J., and Krueger, A. P.,** Long-term biological effects of air ions and D.C. electric fields on Namru mice: first year report, *Int. J. Biometeorol.,* 29, 253, 1985.
27. **Kellogg, E. W., Yost, M. G., Reed, E. J., and Madin, S. H.,** Long-term biological effects of air ions and D.C. electric fields on Namru mice: second year report, *Int. J. Biometeorol.,* 29, 269, 1985.
28. **Krueger, A. P. and Kotoka, S.,** The effects of air ions on brain levels of serotonin in mice, *Int. J. Biometeorol.,* 13, 25, 1969.
29. **Gilbert, G. O.,** Effect of negative air ions upon emotionality and brain serotonin levels in isolated rats, *Int. J. Biometeorol.,* 17, 267, 1973.
30. **Diamond, M. C., Connor, J. R., Orenberg, E. K., Bissell, M., Yost, M., and Krueger, A.,** Environmental influences on serotonin and cyclic nucleotides in rat cerebral cortex, *Science,* 210, 652, 1980.
31. **Quay, W. B.,** Differences in circadian rhythms in 5-hydroxytryptamine according to brain region, *Am. J. Physiol.,* 215, 1448, 1968.
32. **Bailey, W. H., Charry, J. M., Weiss, J. M., Cardle, K., and Shapiro, M.,** Regional analysis of biogenic amine turnover in rat brain after exposure to electrically charged air molecules (air ions), Society for Neuroscience, (Abstr.), 1983.
33. **Bailey, W. H. and Charry, J. M.,** Measurement of neurotransmitter release and utilization in selected brain regions of rats exposed to DC electric fields and atmospheric space charge, in *Proc. 23rd Hanford Life Sciences Symp., Interaction of Biological Systems with Static and ELF Electric and Magnetic Fields,* in press, 1987.
34. **Beardwood, C. J., Jordi, P., and Abrahams, A.,** Alterations in rat flexor withdrawal reflex response to a noxious stimulus after exposure to gaseous air ions, in *Proc. 23rd Hanford Life Sciences Symp., Interaction of Biological System with Static and ELF Electric and Magnetic Fields,* in press, 1987.
35. **Wilson, B. W., Anderson, L. E., Hilton, D. I., and Phillips, R. D.,** Chronic exposure to 60-Hz electric fields: effects on pineal function in the rat, *Bioelectromagnetics,* 2, 371, 1981.
36. **Wilson, B. W., Anderson, L. E., Hilton, D. I., and Phillips, R. D.,** Erratum, *Bioelectromagnetics,* 4, 293, 1983.
37. **Bachman, C. H., McDonald, R. D., and Lorenz, P. J.,** Some physiological effects of measured air ions, *Int. J. Biometeorol.,* 9, 127, 1965.
38. **Knoll, M., Rheinstein, J., Leonard, G. F., and Highberg, P. W.,** Influence of light atmospheric ions on human visual reaction time, *I. R. E. Trans. Biol. Med. Electron.,* 8, 239, 1961.

Chapter 8

HYPOTHETICAL NEURAL SUBSTRATES FOR BIOLOGICAL RESPONSES TO AIR IONS

Robert Kavet

TABLE OF CONTENTS

I. INTRODUCTION

"The search for pulmonary chemoreceptors in the airways and blood vessels has been long and exhaustive owing to the teleologic attractiveness of sensors that would 'taste' the blood and 'smell' the air."[1] This passage from *The Normal Lung* by J. F. Murray addresses primarily chemoreception of the respiratory gases, O_2 and CO_2, as a potential control point to help maintain local ventilation-perfusion ratios. However, if Murray's comment is extended, it raises a question of sensitivity to another gaseous constituent of air, the fraction that carries charge, or, is ionized.

Three factors motivate us to pursue this question: (1) to better understand the basic biological character and environmental interactions of living organisms, (2) to understand whether human activities that alter the charge content of ambient air thereby produce biological effects heretofore unappreciated, and (3) to determine if, in animal or human experimentation, uncontrolled levels of airborne charge might present a hidden source of variability.

The question of whether air ions actually do produce replicable biological effects remains unresolved. As pointed out in recent reviews,[2,3] the literature on ion effects is difficult to interpret because of incomplete reporting or uncertain experimental conditions within many individual studies, and inconsistencies between results from different studies. This chapter explores the coneptual issues facing research on the biological interactions of air ions, and proposes a plausible mechanism of biological action that is consistent with many findings that appear in the literature.

II. AIR IONS

Air ions are charged aggregates of air molecules, and are characterized by their size, chemical composition, net charge, and electrical mobility.[4] Small air ions, the primary subject of biological investigations, consist of singly charged molecules such as H_3O^+ or O_2^- surrounded by a cluster of probably fewer than 10 water molecules; they are about 1 nm in diameter and possess electrical mobilities between 1 and 2 cm^2/V-s. These may combine with other molecules to form medium ions (approximately 10 nm in diameter) or attach to condensation nuclei or aerosol particles to form Langevin ions (up to 0.1 μm in diameter). Collectively, all suspended charged entities are referred to as space charge.

The constitution of ambient space charge is determined by natural phenomena as well as by human activities. Ionizing radiation from the soil, sun, and the cosmos, the turbulent movements of air and water (e.g., waterfalls), and other phenomena such as lightning, are the primary sources of naturally formed air ions. Human contributions to the balance arise from (1) combustion processes that simultaneously generate particles and ions; such particles also tend to subsequently scavenge ions; (2) indoor environments outfitted with synthetic decor and supplied with artificial ventilation, both of which tend to deplete space charge; (3) other situations in which, for example, (a) a corona source such as direct current (DC) overhead transmission line continuously generates plumes of small positive and negative air ions toward the ground; (the description of ion generation by DC lines is pursued in depth further in this volume), or (b) a highly charged surface such as a video display screen depletes its local region of charge.

III. CONCEPTUAL ASPECTS OF AIR ION BIOLOGICAL RESEARCH

Space charge is but one of a myriad of environmental factors that have accompanied the development of human life. For the preponderance of these, the species is equipped with responsive or adaptive systems that either maintain homeostasis or enhance the probability of survival. For example, repair enzymes in the nucleus restore DNA nucleotide sequences disrupted by random "hits" from background ionizing radiation; peripheral blood vessels constrict or dilate appropriately in response to ambient temperature changes; immunocompetent cells recognize foreign matter, such as inhaled bacteria, and participate in its inactivation and clearance; hematocrit and hemoglobin concentrations increase in response to chronic hypoxia as when an individual moves to high altitude. However, we are unaware of afferent systems in man that are sensitive to fluctuations in ambient space charge concentrations.

One frequently advanced argument that dismisses the potential of biological response to air ions posits that they are not present in sufficient quantities to be chemically or electrically effective at biological targets. Dilution of small air ions in air (which contains about 2.7×10^{19} gas molecules per cc at standard conditions) ranges from 10^{-12} to 10^{-11} ppm in urban air to 10^{-11} to 10^{-10} ppm in clean rural air to up to 2×10^{-8} ppm beneath a DC transmission line.[5,6] By comparison, relfex bronchoconstriction mediated by irritant receptors in human airways requires exposure to 1 to 10 ppm SO_2 or 0.01 to 1 ppm O_3;[7,8] olfactory thresholds for these two pollutant gases are an order of magnitude lower, i.e., at least 5×10^5 times higher than the highest possible ion concentrations.[9]

Similar comparisons can be made from an electrical perspective. Air ions in motion produce currents in objects that intercept the ion flux. Under clean air background conditions the entire body collects about 0.01 nA.[10] Worst-case current induction in a person occurs beneath a DC transmission line at the point of maximum ion flow where the body collects less than 3 μA.[11] Such worst-case currents produce resistive heating in tissue approximately 10^{12} times lower than average basal metabolic heat production. The electric fields (approximately 50 μV/m) generated in the CNS by these currents are over 10^6 times less than those required to trigger an action potential and are approximately 10^5 times weaker in magnitude than extracellular EEG signals.[12] From the point of view of conscious sensory stimulation, current densities induced by ions, at their maximum, are at least 10^4 times lower than the minimal DC current densities required for perception by a person.[13]

Such arguments, however, fail to acknowledge the exquisite chemical and electrical sensitivities that have been studied in different biological systems. In this context, Krueger and Reed remind us of the male silkworm's responsiveness to female attractant pheromone when provided at a concentration of only 200 molecules per cm^3 of air and also of the retina's sensitivity to a single photon of light.[14] A recent study reports a sex pheromone produced by *Streptococcus faecalis* active at 4×10^{-11} *M* according to which the authors feel that "only a few molecules" may be necessary to elicit a mating response in the recipient bacterium.[15] In the human kidney, aldosterone and antidiuretic hormone are able to contribute to sodium and water conservation at concentrations of roughly 2×10^{-10} *M* and 10^{-12} *M*;[16,17] these dilutions in aqueous media are not far from the atmospheric dilutions of air ions described earlier.

Similarly, electroreception of extremely small electric fields has been demonstrated in several species of elasmobranch fishes (sharks, skates, and rays) equipped with electric organs responsive to less than 1 μV/m in ambient waters;[18] these signals aid in navigation and precise localization of prey. Bees, pigeons, and certain bacterial species are able to use the earth's magnetic field (approximately 0.5 G) as a reference for forage, navigational, and ritualistic behaviors by virtue of magentosensitive organs.[19-21] Finally, research is now emerging showing that low-level electric signals may interact directly with mammalian brain tissue. In this case no special structures or organs are described; the glycocalyx on the cell surface itself is thought to demodulate the field and trigger transmission of a signal through the plasma membrane via protein molecules.[22]

The aforementioned facts concerning ion effects — namely, lack of a defined receptor and low ambient concentrations — are compounded by difficulties related to identifying

Table 1
CONCEPTUAL DIFFICULTIES IN AIR ION RESEARCH

- Lack of a defined biological sensor or receptor
- Ambient levels of air ions "too small" to be chemically or electrically effective at biological targets
- Lack of reliable markers of exposure and/or effect

markers which faithfully and quantitatively reflect both ion exposure and its consequences (Table 1). Briefly stated, ions do not leave a clear "footprint" in an exposed subject. Many other areas of environmental health science are not burdened to this extent. For example, metabolites of polyaromatic hydrocarbons such as benzo(a)pyrene and l-nitropyrene combine with DNA to form measureable adducts, which reflect target organ exposure and possibly mutagenesis or carcinogenesis; heavy metals such as lead and mercury can be localized in the tissues they effect and body burdens of these substances are frquently possible to assess by sampling blood, hair, or fingernails; carbon monoxide exposure and associated impairment of oxygen delivery to tissues is conveniently evaluated with carboxyhemoglobin measurements; UV exposure affects skin cells by producing pyrimidine dimers, a marker used in association with UV-induced skin cancers.

The examples of biological markers of exposure and/or effect just discussed enjoy two important advantages. First, they are relatively specific to individual environmental agents, and second, they can be quantitatively related to ambient exposures through established dosimetric or monitoring techniques. These advantages are not available to those interested in studying the biological interactions of air ions. In terms of exposure, techniques and instrumentation are now available to characterize ambient environments (such as near a DC transmission line) and to produce tightly controlled artifact-free environments for laboratory animals. (The technical issues and principles concerning ambient measurements and laboratory environments are dealt with further in this book.) Major technical and conceptual problems exist, however, with respect to assessing the exposures of "natural" human populations and free-ranging animals. There are no large data bases that provide information on spatial and temporal distribution of ion concentrations, nor has a personal ion dosimeter yet been devised.

From the perspective of biological interactions, significant hurdles must still be overcome with respect to defining an ion's fate in exposed subjects and the specificity of the biological response to exposure. If air ions do have replicable effects on mammals, they are likely initiated either by inhalation or by electrical induction. The discussion here, however, concentrates on potential interactions following inhalation; the plausibility of this choice is developed in Sections VI and VII below. Ions are highly water soluble and, therefore, deposit most likely in the proximal airways. Unlike inhaled particles which can be precisely visualized microscopically or radiologically at intrapulmonary sites, or gases whose acute responses (bronchoconstriction, edema) can be associated with different lung compartments, inhaled ions deposit on the mucous membrane and quickly relinquish their charge leaving behind no recognizable trail. Clearly, innovative approaches are necessary to characterize accurately deposition patterns of inhaled ions and to pinpoint the possible sites of charge transfer.

Against these disadvantages, human and animal ion research has adopted a "black box" orientation in which a variety of parameters has been examined without much attention directed toward receptor mechanisms. These parameters, as recently compiled by Charry,[2] are listed by study in Tables 2 to 5 together with the reported effects.[23-80] One quickly observes that, collectively, these studies include a wide spectrum of behavioral, physiological, and clinical endpoints. Another scan of the tables will reveal that most or all of these endpoints can register effects from numerous influences unrelated to ion exposure including time of day, nutritional status, drug use, or various stresses. The burden on the investigator is to document control of those extraneous factors in the study design, and then, to record

Table 2
HUMAN PHYSIOLOGY AND BEHAVIOR[a]

Physiological or behavioral response	Air ion effects (+)	Air ion effects (−)	Ref.
Mental performance	0	0	23
Pulse and respiration	0	0	
EEG alpha			24
Frequency	—	d	
Amplitude	—	i	
Synchronization	—	i	
Psychomotor Performance			25
Brightness	0	0	
Reaction time	0	0	
Complex perceptual-motor	0	0	
Muscular steadiness	0	0	
Biochemical responses			
Glucose, steroids, and biogenic amines	0	0	
Metabolism	Biopolar High:	i	26
	Low:	d	
Galvanic skin response	i,d	—	27
Reaction time	0,i	—	
Mood	d	—	
Complex mental task	0	0	28
Vigilance	0	0	
Mood	0	0	
Physical endurance	0	0	29
Anxiety	—	d	30
Performance			31
Motor tasks	0	i	
Reaction time	0	d	
Reaction time	i	d	32
Reaction time	i,d	i,d	33
Reaction time	0	0	34
Vigilance	i	i	
Mood	0	0	35
Mood	0	0	36
Heart rate	0	0	
Respiration	0	0	
Skin temperature	0	0	
Muscle tension	0	0	
Skin resistance	0	0,i	
Blood pressure	0	0,d	
Pulse finger volume	0	0,i	
EEG alpha			37
Frequency	d	d	
Relaxation	i	i	
Sleepiness	i	i	
Dry mucosa	i	i	
Reaction time	i	d	38
Critical flicker-fushion	d	i	
Performance			39
Reaction time	—	d	
Rotary pursuit	—	0	
Mood	—	i	
Reaction time	—	d	40
Respiration rate	0	0	41

Table 2 (continued)
HUMAN PHYSIOLOGY AND BEHAVIOR[a]

Physiological or behavioral response	Air ion effects (+)	Air ion effects (−)	Ref.
Pulse rate	0	0	
Blood pressure	0	0	
Oral temperature	0	0	

Note: — = not studied; 0 = no effects; i = increase; d = decrease

[a] Adapted from Charry, J. M., *Environ. Res.*, 34, 351, 1984.

Table 3
ANIMAL PHYSIOLOGY AND BEHAVIOR[a]

Physiological or behavioral response	Air ion effect (+)	Air ion effect (−)	Ref.
Heart rate	i	i	42
Respiration	i	i	
Activity	i	i,d	43
ECG	i,d	i,d	44
Cortical 5-HT	—	d	45
Cortical cyclic AMP	—	d	
Performance	0	i	46
Avoidance learning	d	i	47
Conditioned emotional response	—	d	48
Reactivity	—	d	49
Brain 5-HT	—	d	
Blood 5-HT	i	d	50
Brain 5-HT	d	d	51
EEG hippocampal theta			52
Frequency	d	i	
Amplitude	i	d	
EGG cortex			
Frontal amplitude	i	—	
Occipital amplitude	i	—	
Conditioned emotional response	d,0	d,0	53
Pain-sensitive defense reactions	—	d	54, 55
Pain-sensitive defense reactions	i	—	56
Spontaneous activity	d	i	57
Expt I: Avoidance Learning	d	i	58
Expt II: Memory	d	0	
Expt III: Spontaneous activity	d	i	
EEG slow wave			59
Frequency	d	i	
Amplitude	i	d	
Maze learning/performance	—	i	60

Note: — = not studied; 0 = no effects; i = increase; d = decrease

[a] Adapted from Charry, J. M., *Environ. Res.*, 34, 351, 1984.

Table 4
CLINICAL SYMPTOMS IN HUMANS[a]

Clinical Symptom	Air ion effects (+)	(−)	Ref.
Burn pain	—	d	61
Postoperative discomfort	—	d	62
Bronchial asthma	—	0,d	63
Postoperative discomfort	—	d	64
Emphysema	0	0	65
Respiratory spasticity	i	d	66
Asthmatic bronchial spasticity	i	d	
Irritation syndrome			67
5-HT urine	i	—	
5-HIAA urine	i	—	
Exhaustion syndrome			
NE, EPI urine	d	—	
17-KS, 17-OH urine	d	—	
Hyperthyroidism			
5-HT urine	i	—	
NE, EPI urine	i	—	
17-KS, 17-OH urine	i	—	
Histamine urine	i	—	
Serotonin — migraine	i	—	68
Exhaustion syndrome			69
NE, EPI urine	0	—	
Irritation syndrome			70
5-HT urine	i	d	
5-HIAA urine	i	d	
Irritation syndrome			71
5-HT urine	i	d	
5-HIAA urine	i	d	
EEG			72
Alpha	—	i	
Synchronization	—	i	
5-HT, 5-HIAA			
Weather-sensitive	—	d	
Normal	—	0	
NE			
Weather sensitive	—	i	
Normal	—	0	
Headache	i	0	73
Dizziness	i	0	
Sore throat	i	0	
Smarting eyes	0	i	
Congested throat	0	i	

Note: — = not studied; 0 = no effects; i = increase; d = decrease

[a] Adapted from Charry, J. M., *Environ. Res.*, 34, 351, 1984.

as fully as possible the environmental conditions under which exposure occurred and data were gathered. Unfortunately, as also critically reviewed by Charry,[2] requisite control was not exercised in the large majority of published studies leaving a situation in which putative ion effects have been described in terms of parameters potentially lacking in specificity to ion exposure.

Table 5
CLINICAL SYMPTOMS IN ANIMALS[a]

Clinical symptom		Air ion effect (+)	Air ion effect (−)	Ref.
Ulceration		—	d	74
Embryonic development				75
Differentiation		0	0	
Course of pregnancy		0	0	
Gestation time		0	0	
Neonatal development				
Survival		0	d	
Growth		d	d	
Chronic respiratory disease		0	i	
Postweaning development		0	i	
Mortality: Coccidioidomycosis		i	—	76
Mortality				77
Pneumonia		i	—	
Influenza		i	—	
Mortality: influenza		—	0,d	78
Mortality: influenza	Low	i	i	79
	High	—	d	
Mortality: influenza	Low	0	0	80
	High	0	0	

Note: — = not studied; 0 = no effects; i = increase; d = decrease

[a] Adapted from Charry, J. M., *Environ. Res.*, 34, 351, 1984.

IV. GENERATING A HYPOTHESIS

Perhaps the best approach to developing a perspective on the potential of air ions to produce biological effects is to first seek common threads in the literature, temporarily overlooking the deficiencies of individual studies and more heavily concentrating on the "weight of evidence". Such an exercise leads one to aggregate studies which differ with respect to species, housing conditions, ion generation, exposure protocols, and bioassay techniques. Other potentially important environmental and experimental factors such as contaminant oxidant gas concentrations, electric fields, and subject grounding are often uncontrolled or unreported. Thus, for obvious reasons, an approach seeking common threads in the ion literature is liberal and somewhat risky, but at least offers the opportunity to frame testable hypotheses. The changes most consistently associated with human or rodent exposure to air ions involve (1) serotonin levels in the periphery (blood or urine) and in the brain, and (2) a variety of behavioral indicators including mood, learning, performance, reaction time, and perception of pain. In general, positive ion exposures are reportedly followed by increases in peripheral serotonin (or its metabolite) and behavioral decrements such as diminished learning and increased reaction time. Negative ion exposures "produce" the opposite outcomes. Paradoxically, both polarities have been associated with decreased brain serotonin.[49] In the studies reporting these results, ion concentrations span three orders of magnitude (10^3 to 10^6 ions/cc), and exposure durations range from less than an hour to months. However, insufficient information exists to allow quantitative generalizations regarding these putative effects as functions of ion concentration or time; it is also not possible to comment on temporal relationships between central and peripheral responses.

The next step is to consider how such effects may occur. An adequate hypothesis (Table 6) must account for ion contact with responsive sites, explain potential peripheral effects

Table 6
CRITERIA FOR AIR ION HYPOTHESIS

- Provide for ion contact with responsive site(s)
- Account for both peripheral and central effects
- Allow for factors that potentially modify responses
  - Inherent (species, sex, age)
  - Induced (diet, L/D cycle)
- Provide for possibility of artifact in previous positive studies

such as changes in blood serotonin levels, and also account for CNS effects as reflected in biochemical or behavioral measures mentioned above. An acceptable proposal should also provide latitude for interaction with factors which could influence response. These may relate to inherent characteristics such as species, strain, sex, and age, or to experimenter manipulation such as drug treatments, dietary deficiencies, light-dark cycles, or changes in respired gas mixtures. Finally, an hypothesis for the biological action of air ions should provide for the possibility that previous studies reported artifact-induced rather than ion-induced effects. Much of the chronicled air ion biological laboratory research was conducted with an incomplete appreciation for the details of ion exposure technology. Thus, artifacts such as oxidant gases, cogenerated with ions, may have consistently in the past induced biological changes attributed to ions.

V. AIR IONS AND SEROTONIN

Whatever the cause, effects on serotonin levels in blood and brain could certainly be linked to a broad range of behavioral and physiological effects (including many of those listed in Tables 2 to 5).[81] In the periphery, serotonin has known constrictor activities on smooth muscle of airways, blood vessels, and alimentary tract; other actions include stimulation of afferent nerve endings and aggregation of platelets. Serotonin does not cross the blood-brain barrier, and so the CNS and the periphery can be considered, for present purposes, as separate compartments. Peripheral serotonin is most abundant in the enterochormaffin cells of the gut (whose function is obscure) and blood-borne serotonin is mainly confined to platelets, which are endowed with a high affinity serotonin uptake mechanism. In addition, free serotonin in plasma is efficiently removed from the blood by pulmonary endothelial cells and metabolized (by monoamine oxidase, discussed below) to 5-hydroxyindoleacetic acid (5-HIAA) which is then excreted in the urine.

In the CNS, neurons containing serotonin are found in the raphe nuclei located along or near the midline from the medulla to the mesencephalon.[81] Axons from these nuclei project rostrally to the limbic system, thalamus and cortex, laterally into the reticular formation of the brain stem, and caudally into the lower brain stem and spinal cord; the cerebellum also receives innervation. The complex array of serotonin-mediated activities in the CNS is summarized as follows: "the evidence suggests that one important general role of the tryptaminergic raphe neurons may be to dampen overreactiveness to various stimuli (external and internal) involving, for example, auditory, visual, olfactory, nociceptive, and other signals affecting social and adaptive behavior (including learning) and involving such parameters as sleep, sexual behavior, aggressiveness, motor activity, perception (including pain), and mood." Furthermore, "tryptaminergic neurons are also involved in other simpler functions such as temperature regulation, neuroendocrine control (regulation of release of hypophysiotrophic hormones) and extrapyramidal activity."[82]

Krueger has proposed that ions exert their action by affecting monoamine oxidase (MAO), an intra-mitochondrial enzyme with a role in the degradative pathway of several amines including serotonin, norepinephrine, and dopamine. MAO is widely distributed throughout the body with highest concentrations in liver and kidney but is also present in the brain and lungs.[83] Krueger supports this "serotonin hypothesis of ion action" with findings from his laboratory that administration of iproniazid, an MAO inhibiting drug, mimics exposure to positive ions; likewise, reserpine, a drug that depletes tissues of serotonin (and norepinephrine), simulates effects of negative ions.[84]

Even if ions do produce the biological effects described, the participation of MAO in mediating them is doubtful for several reasons. First, effects on MAO would involve not only serotonin, but also the catecholamines, dopamine, epinephrine, and norepinephrine. This would produce a full range of peripheral effects associated with the sympathetic nervous system and more profound central effects as well. Secondly, inhaled ions have no access to the central nervous system, nor has an explanation yet been provided to describe CNS involvement even by secondary means. It is difficult for two reasons to conceive of inhaled ions even reaching pulmonary MAO in endothelia where circulating serotonin is metabolized: first, inhaled ions probably do not deposit in alveoli because their high diffusion coefficients predict contact with upper airways and their high water solubility would cause them to be scrubbed at the point of contact;[3] second, endothelial MAO is separated from the adjoining alveolus by the mitochondrial membrane, the basal endothelial membrane, a thin layer of connective tissue and the alveolar epithelium; were an ion to reach an alveolus the probability of its penetrating to the enzyme is remote. Furthermore, monoamine oxidase is not the rate limiting step in the clearance of free serotonin from blood perfusing the lung — the endothelial membrane transport system is;[85,86] even complete inhibition of MAO with iproniazid does not inhibit serotonin's uptake, only its metabolism to 5-HIAA.[86] Finally, Krueger's observation of decreased brain serotonin following exposures to either ion polarity is inconsistent with respective stimulation and inhibition of MAO by negative and positive ions.[51]

Other hypotheses have been advanced that attempt to explain how air ions may trigger biological responses.[3,5,87] However, none have completely addressed the requisite criteria outlined above, and none have identified specifically a plausible receptor that could mediate the reported ion-induced phenomena.

VI. NEUROENDOCRINE CELL HYPOTHESIS

A. Proposal

Over 2 decades ago, Vasiliev, a Russian investigator, put forward the idea that "pulmonary interoceptors" participate in nervous reflex reactions to air ions.[88] The following discussion develops the notion of pulmonary receptors and offers a hypothetical possibility for air ion action that is testable and is consistent with the phenomena reported in the literature.

The hypothesis states that in the respiratory epithelium a small minority population of neuroendocrine cells (NEC) could mediate the range of effects putatively associated with exposure to air ions. NEC, already postulated by many to play a role in the chemoreception of inspired air, also satisfy the criteria for a mechanistic hypothesis discussed earlier: they (1) have, in many cases, direct access to respired air, (2) engage in secretory activity through exocytosis of vesicles that have the capacity to store serotonin and dopamine, and (3) have the potential to transmit signals to the central nervous system via autonomic afferents.

B. Neuroendocrine Cells

1. Description

Neuroendocrine cells of the lung have been described in over 20 species,[89,90] including man, across age groups in some species from fetus to adult. Functional studies of NEC are thus far lacking owing to their sparse distribution in the lung and NEC have been characterized mainly on the basis of a variety of histologic and cytochemical techniques. NEC have acquired various names, including more prominently in the contemporary literature, argyrophilic,[91] small granule,[92] APUD (amine precursor uptake and decarboxylation),[93] Kultschitsky or K cells,[94,95] and Feyrter cells for the man who in 1938 first described endocrinic epithelial cell types.[96-98] Cells with similar properties have been comparatively studied in the gut, thyroid gland, and adrenal medulla;[92,99] in addition, NEC clusters in the lung have been likened to cells of the carotid body and taste buds,[89] both recognized as chemoreceptive structures. Because of diversity related to species, age, and investigator technique, we present a "composite" description of NEC for current purposes and refer the reader to a more complete review literature.[89,90,100]

NEC occur as single cells, referred to here for convenience as K cells, or in clusters or organoids called neuroepithelial bodies (NEB) consisting of from 2 to perhaps as many as 100 cells (Plate 1*).[101] The cells themselves are heterogeneous and have been classified by a number of investigators.[90] Although K cells and NEB differ with respect to distribution in the respiratory tree and association with other constituents of the lung such as nerves, smooth muscle, microvasculature, and other epithelial cells,[102] no major qualitative distinctions between K cells and individual member cells of NEB are reported.

NEC are located at all levels of the pulmonary airway epithelium. They contact the basement membrane and are frequently reported extending to the lumen.[97,103-106] At the luminal surface, they exhibit microvilli and at lateral surfaces form complex interdigitations with neighboring cells. In the cytoplasm, usually in the basal region, neuroendocrine cells are rich with heterogeneous populations of dense core vesicles (DCV) on the order of 100 to 200 nm in diameter.[90] DCV either intrinsically contain serotonin and dopamine or incorporate them, respectively, following administration of 5-hydroxytryptophan and L-DOPA (L-dihydroxyphenylalanine) (Plate 2).[99,104,107] This latter capability has led to their classification as amine precursor uptake and decarboxylation (APUD) cells as described by Pearse.[108] DCV also contain polypeptide hormones such as calcitonin, bombesin, and leu-enkephalin.[109]

2. Function and Innervation

NEC exocytose DCV contents at their basal surface and this secretory activity is stimulated by environmental alterations that include hypoxia and hypercapnea.[110] Such activity confers upon NEC a possible paracrine role whereby their amine and polypeptide secretions may alter their neighboring environment in response to changing ambient conditions. Through such secretory action NEC could influence a variety of cellular entities such as ciliated, goblet, and Clara cells in the respiratory epithelium, or seromucous glands and smooth muscle cells underlying the epithelium. Secretions or their metabolites may also gain access to the circulation so that they are measurable in serum or urine.

Neuroepithelial bodies (NEB) are innervated with both afferent and efferent terminals, the majority appearing to arise from the vagus nerve.[100,103,106,111-113] The presence of synaptic contacts between NEB and these terminals strongly suggests that the NEB secretory products serve as neurotransmitters in addition to their role as paracrine factors (Plate 3).[100] NEB are morphologically similar to the carotid body, well characterized for chemoreception of blood gases. As such, one may speculate that NEB are chemoreceptive organoids sensitive to changes in the gaseous composition of respired air, and as such, help fine-tune local ventilation-perfusion ratios in the lung. Lauweryns and colleagues[114,115] have shown that hypoxia-induced NEB secretion is related to airway rather than blood gas composition, and the response is restricted to the portion of lung under hypoxic stress. NEB may thus represent the long sought chemoreceptive control mechanism (mentioned at the beginning of the chapter) that operates locally through secretion and centrally through neural communication with the CNS. No studies of NEB following air ion exposure have been attempted.

* Plates 1-3 appear following page 174.

C. Neural Pathways

Afferents from pulmonary sensory endings, such as stretch and irritant receptors, enter the medulla of the brain stem via cranial nerve X (the vagus).[116] The fibers terminate in the solitary nucleus of the medulla, a region that also receives sensory information from the carotid body via cranial nerve IX, the glossopharyngeal nerve.[117] Incoming signals relay in the solitary nucleus to other centers in the medulla, pons, hypothalamus, and thalamus concerned with coordinated motor control of respiratory muscle activities, e.g., breathing, coughing.[118,119] The autonomic reflex arc responsible for smooth muscle tone and glandular activity in airways is orchestrated in the hypothalamus and transmitted on the vagus back to the lung on efferents originating in the dorsal motor nucleus;[1,120] (this nucleus also receives sensory input on secondary fibers from the solitary nucleus).[121]

The potential involvement of central nervous serotonergic pathways in the reflex arc just described is not clear, although anatomical considerations certainly allow for the possibility of interaction. For example, in the medulla, the nucleus raphe magnus (one of the nuclei mentioned earlier whose cells contain serotonin) projects efferent fibers to both the solitary nucleus and to the dorsal motor nucleus of the vagus.[122] (The nucleus raphe magnus is also primarily responsible for projecting descending fibers to the spinal cord where serotonergic endings produce analgesic effects.[122])

Perhaps the greatest opportunity for interaction of vagal and serotonergic pathways is in the reticular formation of the brain stem. The reticular formation is a network of nerve tracts and nuclei that communicate afferently and efferently with every level of the central nervous system. Axons from nuclei in the reticular formation project both rostrally and caudally while dendrites and other axonal processes arborize horizontally.[123,124] Among the contributions to this meshwork of axons are projections from the raphe nuclei and secondary sensory fibers from the solitary nucleus.[125,126] Such extensively branching neuronal interconnections in the reticular formation enable the central nervous system to integrate sensory input from the spinal cord and cranial nerves with neuronal inputs from higher centers so that responses back toward the periphery may be coordinated with simultaneous rostral transmission to sites such as the hypothalamus, the thalamus, and the cortex.

One key functional component of the reticular formation and one that relies on the action of serotonin is the ascending reticular activation system, a system concerned with states of consciousness and arousal ranging from keen altertness to sleep.[127] These cortical states (which can be monitored by electroencephalography) are effected at diencephalic loci specifically the hypothalamus, subthalamic regions, and the thalamus largely on the basis of afferents originating in the reticular formation of the brain stem.[128] Serotonergic ascending fibers from rostral raphe nuclei provide powerful inhibitory input to the centers concerned with wakefulness and arousal; in fact, destroying or inhibiting these nuclei leads to insomnia.[129]

D. Summary of NEC Hypothesis

The facts and speculations presented here may be woven into the following hypothesis (Figure 1): air ions produce biological effects as a result of primary contact of inhaled ions with the airway luminal surface of neuroendocrine cells. K cells respond to exposure with changes in their secretory activity with the consequences felt locally consistent with the supposition that K cells fulfill a paracrine role. NEB also respond with secretory activity, but in addition, stimulate afferent vagal terminals. Sensory axons terminating in the solitary nucleus of the medulla synapse with secondary afferents which fan out to respiratory centers of the brain stem, and to the hypothalamus, the dorsal motor nucleus of the vagus, and the reticular formation; in the reticular formation, signals may be provided to serotonergic cells of the raphe nuclei which in turn project to every level of the CNS. Efferents within the CNS carrying input to pulmonary-bound parasympathetics converge on the dorsal motor nucleus of the vagus. Axons from the dorsal motor nucleus travel back to the lung on the

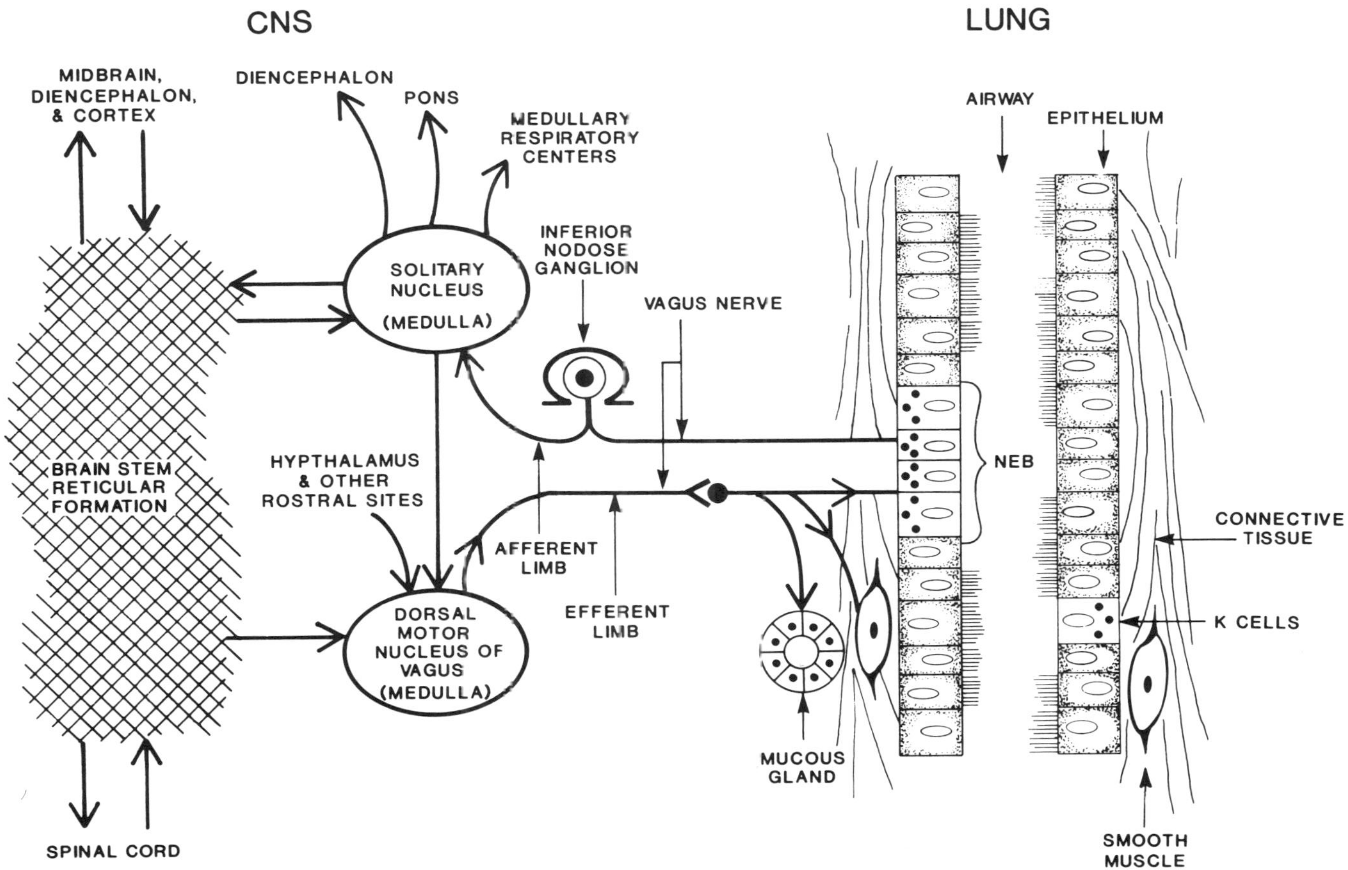

FIGURE 1. Schematic of hypothetical NEB-initiated reflex arc (see text).

Table 7
NEC HYPOTHESIS VS. CRITERIA

Criteria	Neuroendocrine cells
1. Contact with respired ions	1. Airway luminal exposure
2. Responsive to airborne molecules	2. Chemoreceptive properties
3. Peripheral effects	3. Secretory (exocytic) activity
4. Central effects	4. NEB vagally innervated with afferent terminals
5. Modifying factors	5. Species differences and hypoxic effects reported (many others, e.g., age, incompletely studied)
6. Possibility of artifacts in prior studies	6. NEC sensitive to oxidant gases; oxidants also produce confounding effects related to CNS function

vagus completing the arc. Table 7 summarizes the attributes of NEC and matches them to the criteria for an air ion hypothesis discussed earlier; the final two items in the table are dealt with in the following section.

VII. FURTHER CONSIDERATIONS

Activation of these pathways may lead to a range of effects in the periphery and the central nervous system. The brief summary of reported ion effects presented earlier, together with the mechanistic hypothesis just proposed, points to a number of potentially relevant endpoints including, for example: concentrations of neuro-secretory substances (e.g., serotonin, norepinephrine, calcitonin) or their metabolites in lung, blood, urine, and brain; behavioral parameters such as mood, aggressiveness, wakefulness, and pain perception; physiological parameters such as airway resistance and pulmonary arterial pressure that relate to smooth muscle tone or other parameters such as mucus secretion and mucociliary transport in airways and platelet aggregation in blood. Platelets are potentially serotonin-sensitive, and may respond to increased blood serotonin levels with increased aggregation and release of their own stores of serotonin. Platelet aggregation may thus serve as a marker in ion studies.

Within the context of testing the neuroendocrine cell hypothesis, it is necessary to be aware of sources that may introduce significant variability if not controlled by experimental design or at least appreciated. These sources relate to factors in the lung, the central nervous system, and to the treatment of experimental subjects prior to and during ion exposures. (Engineering aspects of ion generation and biological experimentation are vitally important and relevant to this issue and are discussed in other chapters of this volume.) In the lung, the distribution, density, and character of neuroendocrine cells may vary as functions of species and intrapulmonary location; (changes with age, though suggested,[97,98,107,130] are insufficiently investigated). Within cells, the size and content of dense core vesicles vary widely across species;[90] in addition there is some evidence that neuroendocrine cell density varies with location in the lung, decreasing in airways of smaller caliber.[93,101,131,132] The number and distribution of K cells and NEB strongly determine their collective exposure to air ions as the probability for cellular contact with ions diminishes in the deeper lung. Furthermore, the mode and pattern of breathing also determine local ion exposure; for instance, deep mouth breathing produces greater penetration of ions into the lung than shallow nasal breathing.

In the central nervous system, age and sex are important determinants of the number of serotonin and dopamine receptors in human brain.[133] Brain concentrations of these biogenic amines are also subject to diurnal cycles indicating that experimental bioassays also need to be uniform relative to time of day.[134] Moreover, there are references reporting that human sensitivity to ions is dependent upon an individual's baseline emotional or reactive state.[27] These states are related to an individual's ability to cope autonomically with potentially

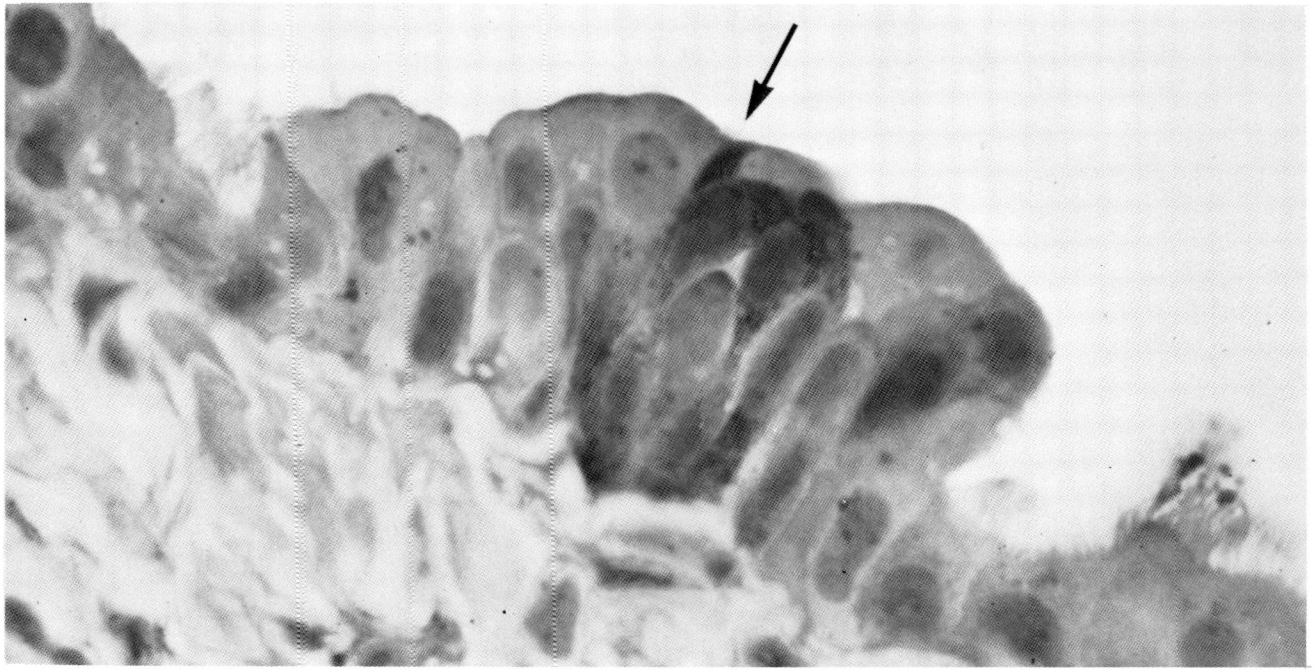

Plate 1. A typical-looking neuroepithelial body from the lung of an adult hamster. It is the darkly stained cluster in the center; the stain is PAS-lead hematoxylin. The lighter adjacent cells are Clara cells and further out are ciliated cells. Smooth muscle underlies the basement membrane. Notice the cellular process from the body reaching to the lumen (arrow). (Magnification × 2400.) (Courtesy of Sorokin, S. P. and Hoyt, R. F., Jr.)

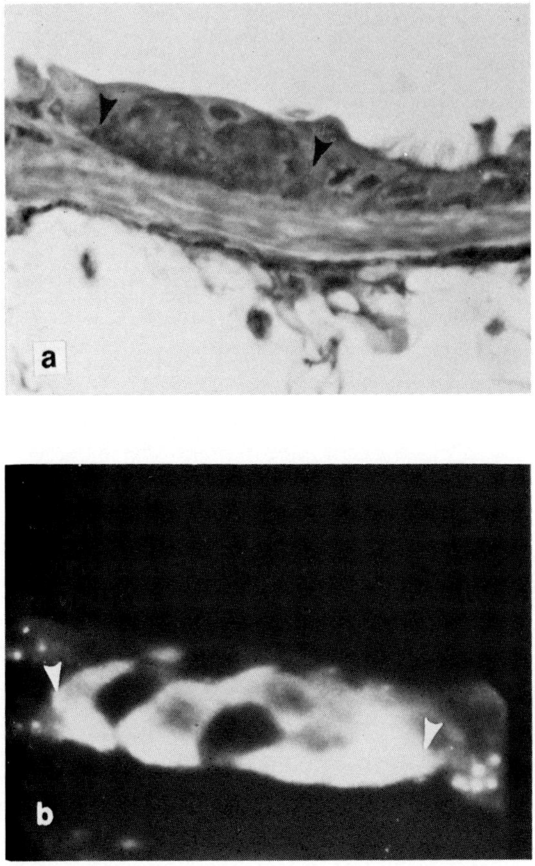

Plate 2. Neuroepithelial body in the lung of a hamster loaded with 5-hydroxytryptophan 30 min prior to fixation in formaldehyde and subsequent embedding in glycol methacrylate. The section is the same for both figures; the notched upper margin of the body and other landmarks help to make this clear (arrowheads mark corresponding locations). (A) Neuroepithelial body after staining with PAS-lead hematoxylin. Cytoplasmic granules are concentrated toward the bases of the cells and are stained well by PAS but only faintly by hematoxylin. (Magnification × 450.) (B) The same body examined by UV microscopy prior to staining. Small-granule cells exhibit yellow fluorescence characteristic of serotonin, which is strongest where the cytoplasmic granules are most abundant. Fluorescence is also visible in lysosomes of epithelial cells to the right and left of the body, but it is bluish-white rather than yellow in color. (Magnification × 800.) (From Sorokin, S. P. and Hoyt, R. F., Jr., *Anat. Rec.*, 192, 245, 1978. With permission.)

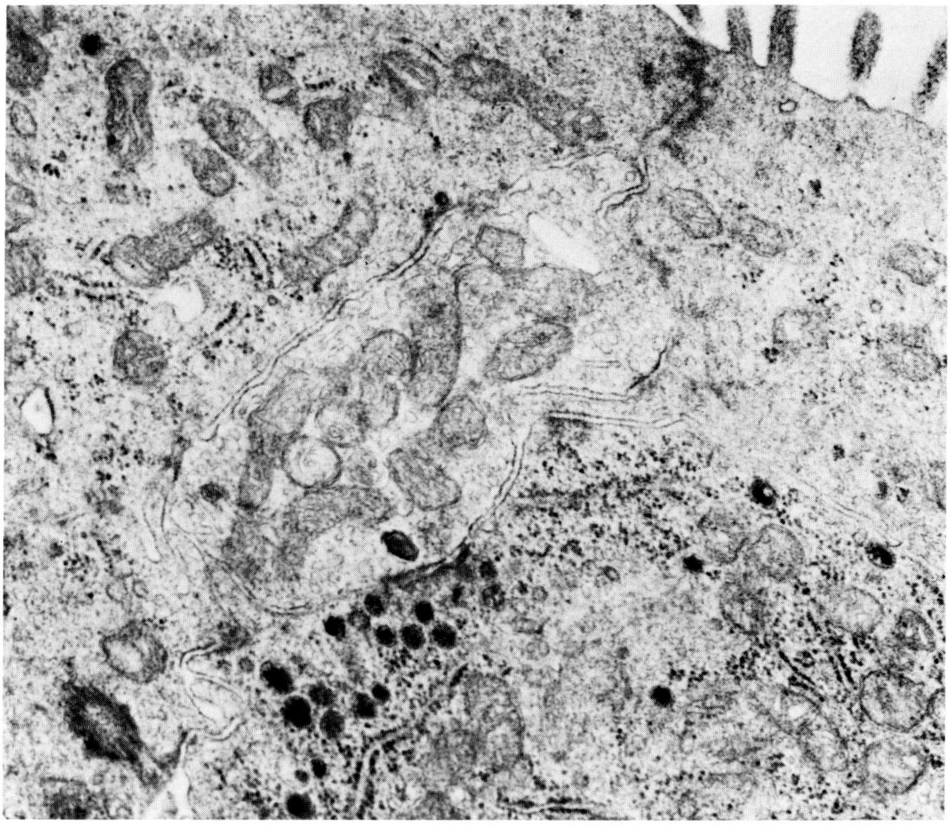

Plate 3. Nerve terminal abutting three small granule cells in a large neuroepithelial body from a 30-day fetal rabbit lung. In this species, bare nerve processes pass between the cells and often terminate close to the surface (right upper corner); these terminals most frequently contain mitochondria and clear synaptic vesicles and are considered to be afferent. Thickenings on the apposed plasmalemmae and a concentration of dense core granules in the vicinity indicate a point of synapse between a small granule cell and the nerve (lower center). (Magnification × 38,000.) (From Sorokin, S. P., Hoyt, R. F. Jr., and Pearsall, A. D., *Am. Res. Resp. Dis.*, 128, S26, 1983. With permission.)

stressful external stimuli. People with low autonomic lability (roughly 25 to 33% of the population) have a distinctly lower reservoir of stress-adaptive capabilities than the remainder of the population, and also appear to be ion-sensitive.[27] Thus, in studies with humans or outbred animals strains, stratification of study subjects according to lability needs to be considered and in inbred strains, emotional/reactive baselines need to be established. Finally, care must be taken to control factors that are unrelated or artifactual to ion exposure. In animals experiments these may be associated with housing, health care, handling, and feeding in addition to other possible environmental sources of variability such as light-dark cycles (known to influence brain serotonin), audible noise, contaminant gases, and DC electric fields. Corona discharge, the method generally used to generate ions, also produces DC electric fields and variable amounts of oxidant gases. Ozone is reported to affect peripheral serotonin levels,[135] as well as a number of behavioral parameters in animals including voluntary activity,[136,137] performance,[138] and pentobarbital-induced sleep time.[139] Furthermore, ozone is also reported to stimulate the proliferation of K-cells in respiratory bronchioles.[95] The effects of DC electric field, if any, are unknown. In human studies some of the precautions corresponding to those taken for laboratory animals are obviously not possible. If air ions do produce biological effects, they are quite likely subtle and thus especially vulnerable to the masking effects introduced by uncontrolled or confounding factors. The validity of experiments testing the neuroendocrine cell hypothesis, or for that matter, any other, lies in the extent to which investigators eliminate unwanted factors.

ACKNOWLEDGMENT

I wish to thank Mary-Ellen Patten for her assistance in the preparation of the manuscript for this chapter.

REFERENCES

1. **Murray, J. F.,** *The Normal Lung, The Basis for Diagnosis and Treatment of Pulmonary Disease,* W. B. Saunders, Philadelphia, 1976, chap. 3.
2. **Charry, J. M.,** Biological effects of small air ions: a review of findings and methods, *Environ. Res.,* 34, 351, 1984.
3. **Bailey, W. H., Bissell, M., Brambl, R. M., Dorn, C. R., Hoppel, W. A., Sheppard, A. R., and Stebbings, J. M.,** *A Health and Safety Evaluation of the ±400 kv DC Powerline,* Minnesota Environmental Quality Board, 1982, chap. 3.
4. **Weast, R. C., Ed.,** *CRC Handbook of Chemistry and Physics,* 60th ed., CRC Press, Boca Raton, Fla., 1979, F-212.
5. **Krueger, A. P. and Sigel, S.,** Small air ions as biologically active agents, in *Biological Effects of Environmental Electromagentism,* Konig, H. L., Krueger, A. P., Lang, S., and Sonning W., Eds., Springer-Verlag, New York, 1981, chap. 8.
6. **Bracken, T. D., Capon, A. S., and Montgomery, D. V.,** Ground level electric fields and ion currents on the Celilo-Sylmar ±400 kVdc Intertie during fair weather, *IEEE Trans. Power Appar. Syst.,* 97, 370, 1978.
7. Committee on Medical and Biologic Effects of Environmental Pollutants, *Ozone and Other Photochemical Oxidants,* National Academy of Sciences, Washington, D.C., 1977, chap. 9.
8. Comittee on Sulfur Oxides, Board on Toxicology and Environmental Health Hazards, *Sulfur Oxides,* National Academy of Sciences, Washington, D.C., 1978, chap. 7.
9. **Goldsmith, J. R. and Friberg, L. T.,** Effects of air pollution on human health, *Air Pollution,* Vol. 2, *The Effects of Air Pollution,* 3rd ed., Stern, A. C., Ed., Academic Press, New York, 1977, chap. 7.
10. **Johnson, G. B.,** The electrical environment and HVDC transmission lines, *Conf. Environ. Ions and Related Biological Effects,* Charry, J. M., Ed., American Institute of Medical Climatology, Philadelphia, 1982, 66.

11. *Transmission Line Reference Book-HVDC to ±600 kV*, Electric Power Research Institute, Palo Alto, Calif., 1977, 74.
12. **Elul, R.,** Dipoles of spontaneous activity in the cerebral cortex, *Exp. Neurol.*, 6, 285, 1962.
13. *Transmission Line Reference Book-HVDC to ±600 kV*, Electric Power Research Institute, Palo Alto, Calif., 1977, 75.
14. **Krueger, A. P. and Reed, E. J.,** Biological impact of small air ions, *Science,* 193, 1209, 1976.
15. **Suzuki, A., Mori, M., Sakagami, Y., Isogai, A., Fujino, M., Kitada, C., Craig, R. A., and Clewell, D. B.,** Isolation and structure of bacterial sex pheromone, cPD1, *Science,* 226, 849, 1984.
16. **Guyton, A. C.,** *Textbook of Medical Physiology,* 6th ed., W. B. Saunders, Philadelphia, 1981, chap. 75.
17. **Guyton, A. C.,** *Textbook of Medical Physiology,* 6th ed., W. B. Saunders, Philadelphia, 1981, chap. 77.
18. **Kalmijn, A. J.,** Electric and magnetic detection in elasmobranch fishes, *Science,* 218, 916, 1982.
19. **Gould, J. L., Kirschvink, J. L., and Deffeyes, K. S.,** Bees have magnetic remanence, *Science,* 201, 1026, 1978.
20. **Walcott, C., Gould, J. L., and Kirschvink, J. L.,** Pigeons have magnets, *Science,* 205, 1027, 1979.
21. **Frankel, R. B., Blakemore, R. P., and Wolfe, R. S.,** Magnetite in freshwater magnetotactic bacteria, *Science,* 203, 1355, 1979.
22. **Lawrence, A. F. and Adey, W. R.,** Nonlinear wave mechanisms in interactions between excitable tissue and electromagnetic fields, *Neurol. Res.*, 4, 115, 1982.
23. **Albrechtsen, O., Clausen, V., Christensen, F. G., Jensen, J. G., and Moller T.,** The influence of small atmospheric ions on human well-being and mental performance, *Int. J. Biometeorol.*, 22, 249, 1978.
24. **Assael, M., Pfeifer, Y., and Sulman, F. G.,** Influence of artificial air ionization on human electroencephalogram, *Int. J. Biometeorol.*, 18, 306, 1974.
25. **Barron, C. I. and Dreher, J. J.,** Effects of electric fields and negative ion concentrations on test pilots, *Aeorsp. Med.*, 35, 20, 1964.
26. **Brandt, A. D.,** The influence of atmospheric ionization upon the human organism, *J. Ind. Hyg.*, 15, 353, 1933.
27. **Charry, J. M. and Hawkinshire, F. B. W.,** Effects of atmospheric electricity of some substrates of disordered social behavior, *J. Personality Soc. Psychol.*, 41, 185, 1981.
28. **Chiles, W. D., Cleveland, J. M., and Fox, R. E.,** A study of the effects of ionized air on behavior, *WADD Technical Report,* 60, 1960.
29. **DeVries, H. A. and Klafs, C. E.,** Ergogenic effects of breathing artificially ionized air, *J. Sports Med. Phys. Fitness,* 5, 7, 1965.
30. **Forward, J. R. and Alpern, H. P.,** Empirical Evaluation of the Effects of Negative Ion Generators and Electrostatic Collectors on Work-Related Stress Behaviors, Study in Social Services Building, Holland, Mich. July 23 to Sept. 9, 1980.
31. **Halcomb, C. G. and Kirk, R. E.,** Effects of air ionization upon the performance of a vigilance task, *J. Eng. Psychol.*, 4, 120, 1965.
32. **Hawkins, L. H. and Barker, T.,** Air ions and human performance, *Ergonomics,* 21, 273, 1978.
33. **Knoll, M., Rheinstein, J., Leonard, G. F., and Highberg, P. W.,** Influence of light atmospheric ions on human visual reaction time, *Proc. Int. Conf. Ionization of the Air,* Vol. 2, American Institute of Medical Climatology, Philadelphia, 1961, XVIII-1.
34. **McDonald, R. D., Bachman, C. H., and Lorenz, C. J.,** Some psychomotor and physiological tests on humans exposed to air ions, *Aerosp. Med.*, 38, 145, 1967.
35. **McGurk, F. C. J.,** Psychological effects of artificially produced air ions, *Amer. J. Phys. Med.*, 38, 136, 1959.
36. **Monaco, R. P. and Acker, C. W.,** Psychophysiological effects of ionized air on psychiatric patients, *Newsl. Res. Psychol.*, 5, 23, 1963.
37. **Silverman, D. and Kornblueh, I. H.,** Effect of artificial ionization of the air on the electroencephalogram, *Am. J. Phys. Med.*, 36, 352, 1957.
38. **Slote, L.,** An experimental evaluation of man's reaction to an ionized air environment, *Proc. Int. Conf. Ionization of the Air,* Vol 2., American Institute of Medical Climatology, Philadelphia, 1961, XX-1.
39. **Tom, G., Poole, M. F., Galla, J., and Berrier, J.,** The influence of negative air ions on human performance and mood, *Hum. Fact.*, 23, 633, 1981.
40. **Wofford, J. C.,** Negative ionization: an investigation of behavioral effects, *J. Exp. Psychol.*, 71, 608, 1966.
41. **Yaglou, C. P., Brandt, A. D., and Benjamin, L. C.,** Observations on a group of subjects before during and after exposure to ionized air, *J. Ind. Hyg.*, 15, 341, 1933.
42. **Bachman, C. H., McDonald, R. D., and Lorenz, P. J.,** Some physiological effects of measured air ions, *Int. J. Biometeorol.*, 9, 127, 1965.
43. **Bachman, C. H., McDonald, R. D., and Lorenz, P. J.,** Some effects of air ions on the activity of rats, *Int. J. Biometeorol.*, 10, 39, 1966.

44. **Bachman, C. H., McDonald, R. D., and Lorenz, P. J.,** Peak changes in electrocardiograms of rats exposed to ionized air, *Int. J. Biometerol.*, 10, 101, 1966.
45. **Diamond, M. C., Connor, J. R., Orenberg, E. K., Bissel, M., Yost, M., and Krueger, A.,** Environmental influences on serotonin and cyclic nucleotides in rat cerebral cortex, *Science*, 210, 625, 1980.
46. **Duffee, R. A., and Koontz, R. H.,** Behavioral effects of ionized air on rats, *Psychophysiology*, 1, 347, 1965.
47. **Flakenberg, V. and Kirk, R. E.,** Effects of ionized air on early acquisition of Sidman avoidance behavior by rats, *Psychol. Rep.*, 41, 1071, 1977.
48. **Frey, A. H.,** Modifications of the conditioned emotional response by treatment with small negative air ions, *J. Comp. Physiol. Psychol.*, 63, 121, 1967.
49. **Gilbert, G. O.,** Effect of negative air ions upon emotionality and brain serotonin in isolated rats, *Int. J. Biometeorol.*, 17, 267, 1973.
50. **Krueger, A. P., Andriese, P. C., and Kotaka, S.,** Small air ions: their effect on blood levels of serotonin in terms of modern physical theory, *Int. J. Biometeorol.*, 12, 225, 1968.
51. **Krueger, A. P. and Kotaka, S.,** The effects of air ions on brain levels of serotonin in mice, *Int. J. Biometeorol.*, 13, 25, 1969.
52. **Lambert, J. F., Olivereau, J. M., and Tuong-Ngoc, A.,** Influence of artificial air ionization on the electroencephalogram of the awake rat, *Int. J. Biometeorol.*, 25, 71, 1981.
53. **Nazzaro, J. R., Jackson, D. E., and Perkins, L. E.,** Effects on ionized air on stress behavior, *Med. Res. Eng.*, 6, 25, 1967.
54. **Olivereau, J. M.,** Effect of negative atmospheric ions on thermalgesia in mice, *C.R. Seances Soc. Biol. Ses. Fil.*, 164, 287, 1970.
55. **Olivereau, J. M.,** Effect of negative atmospheric ions on thermalgesia in mice, *C. R. Seances Soc. Biol. Ses. Fil.*, 164, 746, 1970.
56. **Olivereau, J. M.,** Behavior of mice subjected to thermo-algogenic stimulus after treatment with positive atmospheric ions, *C. R. Seances Soc. Biol. Ses. Fil.*, 164, 501, 1970.
57. **Olivereau, J. M.,** Influence of atmospheric ions on the activity of albino rats, *C. R. Seances Soc. Biol. Ses. Fil.*, 164, 956, 1979.
58. **Olivereau, J. M. and Lambert, J. F.,** Effects of air ions on some aspects of learning and memory of rats and mice, *Int. J. Biometeorol.*, 25, 53, 1981.
59. **Olivereau, J. M., Lambert, J. F., and Truong-Ngoc. A.,** Influence of air ions on brain activity induced by electrical stimulation in the rat, *Int. J. Biometeorol.*, 25, 63, 1981.
60. **Terry, R. A., Harden, D. C., and Mayyasi, A. M.,** Effects of negative air ions, noise, sex and age on maze learning in rat, *Int. J. Biometeorol.*, 13, 39, 1969.
61. **David, T. A., Minehart, J. R., and Kornblueh, I. E.,** Polarized air as an adjunct in the treatment of burns, *Am. J. Phys. Med.*, 39, 111, 1960.
62. **David, T. A., Dereau, J. V., Kornblueh, I. H., McGurk, F. C. J., and Minehart, J. R.,** Ionization of the air: the sedating effect of polarized air, in *Biometeorology*, Tromp. S. W., Ed., Pergamon Press, New York, 1962, 484.
63. **Jones, D. P., O'Connor, S. A., Collins, J. V., and Watson, B. W.,** Effect of long-term air treatment on patients with bronchial asthma, *Thorax*, 31, 428, 1976.
64. **Minehart, J. R., David, T. A., McGurk, F. J., and Kornblueh, I. H., with Derau, J. V.,** The effect of artificially ionized air on post-operative discomfort, *Am. J. Phys. Med.*, 40, 56, 1961.
65. **Motley, H. L. and Yanda, R.,** Environmental air pollution, emphysema, and ionized air, *Dis. Chest.*, 50, 343, 1966.
66. **Palti, Y., DeNour, E., and Abrahamov, A.,** The effect of atmospheric ions on the respiratory system of infants, *Pediatrics*, 38, 405, 1966.
67. **Sulman, F. G., Danon, A., Pfeifer, Y., Tal, E., and Weller, C. P.,** Urinalysis of patients suffering from climatic heat stress (Sharav), *Int. J. Biometeorol.*, 14, 45, 1970.
68. **Sulman, F. G.,** Serotonin-migraine in climatic heat stress, its prophylaxis and treatment, *Proc. Int. Headache Symp.*, 1971, 205.
69. **Sulman, F. G., Pfeifer, Y., and Superstine, E.,** Adrenal medullary exhaustion from tropical winds and its management, *Isr. J. Med. Sci.*, 9, 1022, 1973.
70. **Sulman, F. C., Levy, D., Pfeifer, Y., Superstine, E., and Tal, E.,** Airionometry of hot, dry desert winds (Sharav) and treatment with air ions of weather-sensitive subjects, *Int. J. Biometeorol.*, 18, 313, 1974.
71. **Sulman, F. G., Levy, D., Pfeifer, Y., Superstine, E., and Tal, E.,** Effects of the Sharav and Bora on urinary neurohormone excretion in 500 weather-sensitive females, *Int. J. Biometeorol.*, 19, 202, 1975.
72. **Sulman, F. G., Levy, D., Lunkan, L., Pfeifer, Y., and Tal, E.,** Absence of harmful effects of protracted negative air ionization, *Int. J. Biometeorol.*, 22, 53, 1978.
73. **Winsor, T. and Beckett, J. C.,** Biologic effects of ionized air in man, *Am. J. Phys. Med.*, 37, 83, 1958.

74. **Deleanu, M., Fritz, T., and Florea, E.,** Action of ionized air in the treatment of gastro-duodenal ulcers, *Int. J. Biometeorol.*, 9, 161, 1965.
75. **Hinsull, J., Bellamy, D., and Head, E. L.,** Effects of air ions on the neonatal growth of laboratory rats, *Int. J. Biometeorol.*, 25, 323, 1981.
76. **Krueger, A. P. and Levine, H. B.,** The effect of unipolar positively ionized air on the course of coccidioidomycosis in mice, *Int. J. Biometeorol.*, 11, 279, 1967.
77. **Krueger, A. P., Kotaka, S., Reed, E. J., and Turner, S.,** The effects of air ions on bacterial and viral penumonia in mice, *Int. J. Biometeorol.*, 14, 247, 1970.
78. **Krueger, A. P., Kotaka, S., and Reed, E. J.,** The course of experimental influenza on mice maintained in high concentration of small negative air ions, *Int. J. Biometeorol.*, 15, 5, 1971.
79. **Krueger, A. P. and Reed, E. J.,** Effects of the air ion environment on influenza in the mouse, *Int. J. Biometeorol.*, 16, 209, 1972.
80. **Krueger, A. P., Reed, E. J., Day, M. B., and Brook, K. A.,** Further observations on the effect of air ions on influenza in the mouse, *Int. J. Biometeorol.*, 18, 46, 1974.
81. **Douglas, W. W.,** Histamine and 5-hydroxytryptamine (serotonin) and their antagonists, in *Goodman and Gilman's The Pharmacological Basis of Therapeutics,* 6th ed., Gilman, A. G., Goodman, L. S., and Gilman, A. G., Eds., Macmillan, New York, 1980, chap. 26.
82. **Douglas, W. W.,** Histamine and antihistamines: 5-hydroxytryptamine and antagonists, in *The Pharmacological Basis of Therapeutics,* 5th ed., Goodman, L. S., and Gilman, A. G., Eds., Macmillan, New York, 1975, chap. 29.
83. **Mayer, S. E.,** Neurohumoral transmission and the autonomic nervous system, in *Goodman and Gilman's The Pharmacological Basis of Therapeutics,* 6th ed., Gilman, A. G., Goodman, L. S., and Gilman, Eds., Macmillan, New York, 1980, chap. 4.
84. **Krueger, A. P. and Smith, R. F.,** The biological mechanisms of air ion action: I. 5-hydroxytryptamine as the endogenous mediator of positive air ion effects on the mammalian trachea, *J. Gen. Physiol.*, 43, 533, 1960.
85. **Block, E. R.,** Early metabolic changes in response to lung injury: extrapolation from animals to humans, *J. Toxicol. Environ. Health,* 13, 369, 1984.
86. **Fisher, A. B., Block, E. R., and Pietra, G.,** Environmental influences on uptake of serotonin and other amines, *Environ. Health Perspect.*, 35, 191, 1980.
87. **Kellogg, E. W., III,** Air ions: their possible biological significance and effects, *J. Bioelectr.*, 3, 119, 1984.
88. **Vasiliev, L. L.,** The physiological functions of aerions, *Am. J. Physiol. Med.*, 39, 124, 1960.
89. **DiAugustine, R. P. and Sonstegard, K. S.,** Neuroendocrine-like (small granule) epithelial cells of the lung, *Environ. Health Perspect.*, 55, 271, 1984.
90. **Becker, K. L. and Gazdar, A. F.,** The pulmonary endocrine cell and the tumors to which it gives rise, in *Comparative Respiratory Tract Carcinogenesis,* Vol. 2, Reznik-Schuller, M. M., Ed., CRC Press, Boca Raton, Fla., 1983, chap. 7.
91. **Taylor, W.,** Pulmonary argyrophil cells at high altitude, *J. Pathol.*, 122, 137, 1977.
92. **Sorokin, S. P. and Hoyt, R. F., Jr.,** PAS-lead hematoxylin as a stain for small-granule endocrine cell populations in the lungs, other pharyngeal derivatives and the gut, *Anat. Rec.*, 192, 245, 1978.
93. **Kleinerman, J., Marchevsky, A. M., and Thornton, J.,** Quantitative studies of APUD cells in airways of rats: the effects of diethylnitrosamine and NO2, *Am. Rev. Resp. Dis.*, 124, 458, 1981.
94. **Bensch, K. G., Gordon, G. G., and Miller, L. R.,** Studies on the bronchial counterpart of the Kultschitzky (argentaffin) cell and innervation of bronchial glands, *J. Ultra. Res.*, 12, 668, 1965.
95. **Castleman, W. L., Dungworth, D. L., Schwartz, L. W., and Tyler, W. S.,** Acute respiratory bronchiolitis: an ultrastructural and autoradiographic study of epithelial cell injury and renewal in rhesus monkeys exposed to ozone, *Am. J. Pathol.*, 98, 811, 1980.
96. **Feyrter, F.,** *Über Diffuse Endokrine Epitheliale Organic,* J. A. Barth, Leipzig, 1938.
97. **Moosavi, H., Smith, P., and Heath, D.,** The Feyrter cell in hypoxia, *Thorax,* 28, 729, 1973.
98. **Hernandez-Vasquez, A., Will, J. A., and Quay, W. B.,** Quantitative characteristics of the Feyrter (APUD) cells of the neonatal rabbit lung in normoxia and chronic hypoxia, *Thorax,* 32, 449, 1977.
99. **Hoyt, R. F., Jr., Sorokin, S. P., and Bartlett, R. A.,** A simple flourescence method for serotonin-containing endocrine cells in plastic-embedded lung, gut, and thyroid gland, *J. Histochem. Cytochem.*, 27, 721, 1979.
100. **Sorokin, S. P., Hoyt, R. F., Jr., and Pearsall, A. D.,** Comparative biology of small granule cells and neuroepithelial bodies in the respiratory system: short review, *Am. Res. Resp. Dis.*, 128, S26, 1983.
101. **Hoyt, R. F., Jr., Sorokin, S. P., and Feldman, H.,** Small-granule (neuro)endocrine cells in the infracardiac lobe of a hamster lung. Number, subtypes, and distribution, *Exp. Lung Res.*, 3, 273, 1982.
102. **Hoyt, R. F., Jr., Feldman, H., and Sorokin, S. P.,** Neuroepithelial bodies (NEB) and solitary endocrine cells in the hamster lung, *Exp. Lung Res.*, 3, 299, 1982.

103. **Hung, K. S. and Loosli, C. G.,** Bronchiloar neuro-epithelial bodies in the neonatal mouse lungs, *Am. J. Anat.,* 140, 191, 1974.
104. **Lauweryns, J. M., Cokelaere, M., and Theunynck, P.,** Serotonin producing neuroepithelial bodies in rabbit respiratory mucosa, *Science,* 180, 410, 1973.
105. **Lauweryns, J. M. and Goddeeris, P.,** Neuroepithelial bodies in the human child and adult lung, *Am. Rev. Respir. Dis.,* 111, 469, 1975.
106. **Stahlman, M. T. and Gray, M. E.,** Ontogeny of neuroendocrine cells in human fetal lung. I. An electron microscope study, *Lab. Invest.,* 51, 449, 1984.
107. **Hage, E.,** Light and electron microscopic characteristics of the various lung endocrine cell types, *Invest. Cell Pathol.,* 3, 345, 1980.
108. **Pearse, A. G. E.,** The cytochemistry and ultrastructure of polypeptide hormone-producing cells of the APUD series and the embryologic, physiologic, and pathologic implications of the concept, *J. Histochem. Cytochem.,* 17, 303, 1969.
109. **Stahlman, M. T., Kasselberg, A. G., Orth, D. N., and Gray, M. E.,** Ontogeny of neuroendocrine cells in human fetal lung. II. An immunohistochemical study, *Lab Invest.,* 52, 52, 1985.
110. **Lauweryns, J. M., Cokelaere, M., Deleersnyder, M., and Liebens, M.,** Intrapulmonary neuro-epithelial bodies in newborn rabbits: influence of hypoxia, hyperoxia, hypercapnea, nicotine, reserpine, L-DOPA, and 5-HTP, *Cell Tissue Res.,* 182, 425, 1977.
111. **Lauweryns, J. M. and Cokelaere, M.,** Hypoxia-sensitive neuro-epithelial bodies: intrapulmonary secretory neuroreceptors, modulated by the CNS, *Z. Zellforsch.,* 145, 521, 1973.
112. **Carabba, V. H., Sorokin, S. P., and Hoyt, R. F., Jr.,** Development of neuroepithelial bodies in intact and cultured lungs of fetal rats, *Am. J. Anat.,* 173, 1, 1985.
113. **Lauweryns, J. M. and Van Lommel, A.,** The intrapulmonary neuroepithelial bodies after vagotomy: demonstration of their sensory neuroceptor-like innervation, *Experientia,* 39, 1123, 1983.
114. **Lauweryns, J. M., De Bock, V., Guelinckx, P., and Decramer, M.,** Effects of unilateral hypoxia on neuroepithelial bodies in rabbit lungs, *J. Appl. Physiol.: Respir., Environ. Exercise Physiol.,* 55, 1665, 1983.
115. **Lauweryns, J. M., Cokelaere, M., Lerut, T., and Theunynck, P.,** Cross-circulation studies on the influence of hypoxia and hypoxaemia on neuro-epithelial bodies in young rabbits, *Cell Tissue Res.,* 193, 373, 1978.
116. **Comroe, J. H.,** *Physiology of Respiration,* Year Book Medical Publishers, Chicago, 1965, chap. 8.
117. **Carpenter, M. B. and Sutin, J.,** *Human Neuroanatomy,* 8th ed., Williams & Wilkins, Baltimore, 1983, 339.
118. **Carpenter, M. B. and Sutin, J.,** *Human Neuroanatomy,* 8th ed., Williams & Wilkins, Baltimore, 1983, chap. 11.
119. **Haines, D. E.,** *Neuroanatomy: An Atlas of Structures, Sections, and Systems,* Urban and Schwarzenberg, Baltimore, 1983, 140
120. **Carpenter, M. B. and Sutin, J.,** *Human Neuroanatomy,* 8th ed., Williams & Wilkins, Baltimore, 1983, 571.
121. **Carpenter, M. B. and Sutin, J.,** *Human Neuroanatomy,* 8th ed., Williams & Wilkins, Baltimore, 1983, chap. 11.
122. **Carpenter, M. B. and Sutin, J.,** *Human Neuroanatomy,* 8th ed., Williams & Wilkins, Baltimore, 1983, 408.
123. **Noback, C. R. and Demarest, R. J.,** *The Nervous System: Introduction and Review,* McGraw-Hill, New York, 1972, 90.
124. **Carpenter, M. B. and Sutin, J.,** *Human Neuroanatomy,* 8th ed., Williams & Wilkins, Baltimore, 1983, chap. 11.
125. **Carpenter, M. B. and Sutin, J.,** *Human Neuroanatomy,* 8th ed., Williams & Wilkins, Baltimore, 1983, 297.
126. **Carpenter, M. B. and Sutin, J.,** *Human Neuroanatomy,* 8th ed., Williams & Wilkins, Baltimore, 1983, 441.
127. **Noback, C. R. and Demarest, R. J.,** *The Nervous System: Introduction and Review,* McGraw-Hill, New York, 1972, chap. 19.
128. **Carpenter, M. B. and Sutin, J.,** *Human Neuroanatomy,* 8th ed., Williams & Wilkins, Baltimore, 1983, 442.
129. **Carpenter, M. B. and Sutin, J.,** *Human Neuroanatomy,* 8th ed., Williams & Wilkins, Baltimore, 1983, 409.
130. **Hage, E., Hage, J., and Juel, G.,** Endocrine-like cells of the pulmonary epithelium of the human adult lung, *Cell Tissue Res.,* 178, 39, 1977.
131. **Palisano, J. R. and Kleinerman, J.,** APUD cells and neuroepithelial bodies in hamster lung: methods, quantitation, and response to injury, *Thorax,* 35, 363, 1980.

132. **Marchevsky, A. M., Keller, S., Fogel, J. R., and Kleinerman, J.,** Quantitative studies of argyrophilic APUD cells in airways. III. The effects of sensitization and anaphylactic shock, *Am. Rev. Respir. Dis.,* 129, 477, 1984.
133. **Wong, D. F., Wagner, H. N., Jr., Dannals, R. F., Links, J. M., Frost, J. J., Ravert, H. T., Wilson, A. A., Rosenbaum, A. E., Gjedde, A., Douglass, K. H., Petronis, J. D., Folstein, M. F., Toung, J. K. T., Burns, H. D., and Kuhar, M. J.,** Effects of age on dopamine and serotonin receptors measured by positron tomography in the living human brain, *Science,* 226, 1393, 1984.
134. **Hery, F., Rouer, E., and Glowinski, J.,** Daily variations of serotonin metabolism in the rat brain, *Brain Res.,* 43, 445, 1972.
135. **Veninga, T. S.,** Toxicity of ozone in comparison with ionizing radiation, *Strahlentherapie,* 134, 469, 1967.
136. **Murphy, S. D., Ulrich, C. E., Frankowitz, S. H., and Xintaras, C.,** Altered function in animals inhaling low concentrations of ozone and nitrogen dioxide, *J. Am. Ind. Hyg. Assoc.,* 25, 246, 1964.
137. **Tepper, J. L., Weiss, B., and Cox, C.,** Microanalysis of ozone depression of motor activity, *Toxicol. Appl. Pharmacol.,* 64, 317, 1982.
138. **Weiss, B., Ferin, J., Merigan, W., Stern, S., and Cox, C.,** Modification of rat operant behavior by ozone exposure, *Toxicol. Appl. Pharmacol.,* 58, 244, 1981.
139. **Graham, J. A., Menzel, D. B., Miller, F. J., Illing, J. W., and Gardner, D. E.,** Influence of ozone on pentobarbital-induced sleeping time in mice, rats, and hamsters, *Toxicol. Appl. Pharmacol.,* 61, 64, 1981.

Chapter 9

HISTORY OF AIR ION RESEARCH

Alfred P. Wehner

TABLE OF CONTENTS

Do not follow where the path may lead.
Go instead where there is no path and leave a trail.

This chapter is dedicated to the
pioneers in air ion research.

I. INTRODUCTION

The history of air ion research is the story of a controversial field and the individuals who shaped it. Viewed from the perspective of scientific research in the 1980s, much of the air ion research has scientific shortcomings, especially the earlier work. My personal experience in this field ranges from impressive clinical results with electro-aerosol inhalation therapy in patients with respiratory disorders[1,2] to experiments, conducted with state-of-the-art technology, which failed to affect any of more than 20 blood parameters,[3] although such effects had been reported by many of the early investigators (see below). The dichotomy of results appears to be typical of air ion research, but then, controversy in the form of contradictory or inconsistent results is not unique to air ion research as it is rather common in other branches of biomedical research. A discussion of the scientific merits of past air ion research would exceed the scope of this historical chronicle and is given elsewhere in this book. An extensive, albeit incomplete, list of references is included in this chapter so that the history buff or the scientist interested in specific aspects of some of the earlier work may be able to study the original publications and form his/her own opinion.

As we stand on the shoulders of our predecessors in this field, looking both back and ahead, we continue to be puzzled, and sometimes frustrated, by the complexity and seeming fickleness of the behavior of air ions and their effects on biological systems. Only continuous, scientifically sound research will provide some of the answers which, in turn, almost certainly will raise new questions.

II. THE EARLY DAYS

The effect of climate, weather, and ambient air on the well-being of man can be traced back a couple of millenia. The Greek physician Claudius Galenus (131 to 201 A.D.) empirically recognized the salutory effects of sea air in certain pulmonary diseases and recommended to his patients sea voyages or permanent seaside residence, combined with breathing exercises. This treatment lead to the development of spas and is still widely and successfully used in European medicine. Likewise, the effects of climate and weather on health are well known to modern medical practitioners and patients alike. They manifest themselves in weather-sensitive patients in the form of exacerbations in asthmatics, flare-ups of pain in amputation stumps and rheumatic joints, headaches, mood changes, and various other symptoms. Exactly which component, or combination of components, of the weather or climate is responsible for these phenomena remains to be determined, although air ions appear to be involved at least to some degree.

After his death, the findings of Galenus fell into oblivion. Many centuries later, Franklin (1751)[4] and Dalipard (1752)[5] independently discovered atmospheric electricity. In 1780, the Abbé Bertholon published the first treatise on biological effects of electricity in humans.[6] In 1795, Coulomb demonstrated that air conducts electricity.[7] In 1899, Elster and Geitel showed that molecular and submolecular particles, which they called "ions", were responsible for this phenomenon.[8] In 1901, Czermak postulated that positively charged air ions (in the following also called positive ions) cause deleterious effects in weather-sensitive individuals.[9] Following Czermak, Sokolof[10] and Steffans[11] were among the first of an increasing number of physicians and naturalists to attribute adverse effects to peculiar regional ''winds of ill repute'', such as the Föhn in the Alps. In the 1920s and 1930s, air ion research began to enter its most prolific era with the work of Dessauer[12] in Germany and Tchijevsky[13-23] in the Soviet Union. It is on this era and the following decades on which this chapter will focus.

III. THE ERA OF DESSAUER AND TCHIJEVSKY

Once atmospheric electricity in the form of air ions had been perceived to have various biological effects, several devices were developed to generate positively or negatively charged air ions on demand for clinical and laboratory studies. One such system used induction and corona discharge. Corona discharge usually produces ozone with biological effects of its own. It appears that the early investigators either were not aware of, or ignored, this problem. Dessauer and his colleagues designed an apparatus based on the effect of glowing magnesium or calcium oxide. Dessauer's research activity resulted in the publication of the book *Zehn Jahre Forschung auf dem Physikalisch-Medizinischen Grenzgebiet* (Ten Years of Research in the Physicomedical Border Area), a classic in its field.[12] Stimulated by Dessauer's pioneering work, numerous other investigators began to study biological effects of air ions.

Yaglou et al.,[24] using a Dessauer apparatus, exposed patients to negative and positive air ions. They reported a decrease in blood pressure in 70% of their patients, an increase in 20%, and no effect in 10% after treatment with negatively charged ions. Exposure to positively charged ions reportedly resulted in higher blood pressure in 62% of the patients and in lower blood pressure in 23%. There was no change in 15%.

Edstrom, using a Dessauer ion generator, described a decrease of muscle chronaxie, blood pressure, and basal metabolism rate following exposure to negative ions.[25,26] Opposite results were observed following exposure to positive ions. He reported a decrease in skin temperature in 23 out of 27 patients exposed to negative ions and an increase in 23 cases following exposure to positive ions.[27] Blood pressure reportedly decreased in 75% of hypertensive patients treated with negatively charged ions and increased following exposure to positive ions. Leicher reported to have ''cured'' ten cases of chronic and mucopurulent rhinitis with a Dessauer apparatus and negative ionization.[28] Strasburger and Lampert used the same system to treat 81 patients with suppurative sinusitis and claimed the following results: 41 considerably improved, 20 moderatly improved, and 20 showed no effect.[29]

Wilhelmi and Wilhelmi, using a Dessauer apparatus for the generation of negative air ions, reported a decrease in blood pressure between 15 and 45 mm in 29 hypertensive patients; 17 cases showed subjective improvement. Of 14 hypertensive patients with nephritis, 6 showed decreases between 15 to 40 mm.[30] Fervers, using a Dessauer system, reported a marked decrease in blood sedimentation rates in ion-exposed subjects.[31] Lampert treated migraine, rhinitis, the common cold, nervous sleeplessness, and other diseases using an A.E.G. air ion generator and claimed improvements in his patients.[32] Landsmann, using a corona-discharge type ion generator, treated 60 asthmatics. He observed a decrease in eosinophils and lymphocytes and claimed cessation of asthma attacks in 31 of the treated patients.[33] Achler, using a Wimshurst machine as an air ion generator, investigated the effects of air ions on the respiratory tract of rabbits. His results indicated, according to Achler, that air ions affect the vasomotor nerves of the mucous membranes of the respiratory tract.[34] Baumgarten, using a Küster-Meixner apparatus, described longer survival of ion-treated mice which had been infected with *Salmonella enteritidis* Gärtner.[35]

The late A. L. Tchijevsky, former Director of the Central Laboratory for Ionization in Moscow, can be considered the father of air ion research in the Soviet Union. His activities resulted in numerous publications, including a voluminous, well-referenced book in 1960, entitled, in translation, *Air Ionization, Its Role in the National Economy*.[22] The fact that this book was published by the Publishing House of the State Planning Commission of the USSR in Moscow implies official sanction and recognition of his work by the Soviet government.

Tchijevsky used corona discharge for the generation of air ions. He reported, among other findings, increased activity in rats exposed to air ions. The animals, according to Tchijevsky, showed both accelerated growth and significant weight gains, and they produced healthier and stronger offspring than the control animals. Tchijevsky's clinical trials resulted in improvement in 12 tuberculosis patients as evidenced by considerable weight gain, disappearance of cough and perspiration, and by negative sputum.[13] Tchijevsky also reported the ''cure'' of a tuberculous chimpanzee. With Bobrov he observed a 15% accelerated wound healing time in mice.[20] In other clinical tests he reported decreases in blood pressure of up to 39 mm in hypertensive patients.[21]

Given the difficult access to Russian technical literature — in the 1920s and 1930s more so than today — most of Tchijevski's publications and those of his contemporary colleagues in this field are all but unknown in the West, and their scientific quality and validity is rather difficult to assess.

Rossiisky and Vilenkin,[35a] also using a corona-discharge type ion generator, claimed remarkable results with the therapeutic application of negative ions in 300 influenza cases. The duration of the disease reportedly was reduced from an average of 5 to 8 days to only 2 to 3 days. Kaufmann et al.,[36] like Tchijevsky, reported on the treatment of tuberculosis with air ions. Pachinkov[37] treated bronchial asthma with air ions. Prosorovsky[38] used air ions in the treatment of autonomic and endocrine disorders as well as bronchial asthma. Oserkowsky[39] studied the quantitative relation between chlorophyll and iron in pear leaves. Sapoznikova[40] studied respiration of weed seed in ionized air. Vysotskii[41,42] reported on the effect of air ions on the cotton plant. Zvonitzky and Obrossov[43] reviewed biological effects of high concentrations of air ions.

Japanese air ion researchers were among the most active in this era. Abe reported inhibitory effects of air ions on tumor development.[44,45] He also investigated air ion effects on blood sugar and growth in guinea pigs.[46] Anazawa studied air ion action on blood pressure[47,48] and constipation.[49] Ariizumi described effects on enzyme solutions.[50] Ashiba claimed an antihypertensive effect[51,52] and also reported on air ion effects on the frog heart.[53] Fukui described effects on respiration in animals[54,55] and on diuresis.[56,57] Fukuzumi reported effects on bacterial growth,[58] on the reticuloendothelial system,[59] and on experimental anemia.[60] Honma and his colleagues claimed effects on blood and cardiovascular parameters.[61-63] Horimi and Hasegawa,[64,65] Miyoshi,[66] Nakajima,[67,68] Ogawa,[69] Kusaka and Itano,[70] and Ando[71] used air ions in the treatment of hypertension. Ishidate described effects on various blood parameters,[72,73] as did Karube,[74,75] Murajama,[76] Nakujama,[77] Sakuma,[78-83] Sekino and Nakamura,[84] Suzuki,[85] Takahashi,[86-88] Takasaki,[89] and Tsuchida.[90]

A number of Japanese investigators of this era reported on the therapeutic use of air ions in various conditions other than hypertension. Among them were Matsugaki and Hasuo,[91] Kishioka,[92] Kojima,[93] Matsuhashi,[94] Matsui,[95] Minamide and Isoya,[96] Minamiura,[97,98] Miyoshi,[99-101] Mori,[102] Murakoshi,[103] Ohnishi,[104] Ohta,[105] Takahashi,[106] Togasawa,[107,108] Kimura and Tanigushi,[109] and Kakita.[110]

Other Japanese authors investigated various other biological effects of air ions. Jo reported on the effects of ions on bacterial luminescence.[111] Kimura[112] and Kimura et al.[113] researched effects of air ions, and the effects of air ion depletion on humans in occupied rooms. Ohnishi et al. related environmental hygiene in cities and rooms to air ion concentrations.[114-116] Moritani studied ion effects on bacterial growth[117] and experimental rickets.[118-120] Ohmura[121-123] investigated effects on protein metabolism. Murayama[125] studied effects of ions on the growth of the guinea pig[124] and on blood cholesterol and fatty acids. Ohta[126] described effects on the development and metamorphosis of the toad. Saito[127-130] researched ion effects on chronaxy in the pigeon under various conditions. Sekino[131-133] observed ion effects on phagocytosis, the reticuloendothelial system and blood pressure in the rabbit. Sone[134,135] found effects on experimental hypertension in rabbits. Sinaya[136] claimed effects of air ions on electrophoresis of erythrocytes. Ion effects on experimental scurvy were reported by Mugikura.[137,138] Effects on the silkworm were described by Miura.[139] Yuasa[140,141] examined ion effects on the perfused excised toad heart and Sugano[142] reported on the effects of ions on the growth of the tubercle bacillus.

The Japanese literature of this era presents to Western scientists problems in language and accessibility similar to Russian publications of that time. This makes it difficult to evaluate the quality of the Japanese investigations.

IV. THE ERA OF KRUEGER AND KORNBLUEH

Compared to the 1930s, the 1940s and most of the 1950s are characterized by a paucity of publications in the air ion field, perhaps due to the disruptive effects of World War II and its aftermath. However, during the second half of the 1950s renewed interest in air ions began to emerge, closely associated with the activities of the pioneers of this new era, Krueger and Kornblueh.

Albert Paul Krueger (1902—1982) at the University of California in Berkeley was, without doubt, the most prominent and most respected pioneer in the field of basic air ion research. He could look back at a distinguished career as a medical scientist when, in his 50s, his main interests began to focus on biological effects of air ions. Subsequently, some 50 papers originated from the Air Ion Research Laboratory at the University of California under his investigative leadership, more than 40 of them with Krueger as the principal author.[143-195] His mind as sharp as ever, he was still active in air ion research when death claimed him at thc agc of 80.

Krueger and his co-workers reported statiscally significant and reproducible stimulation of ciliary activity in the trachea of rabbits, mice, guinea pigs, rats, primates, and man exposed to negative ions. Positive air ions, according to Krueger et al., decreased or stopped ciliary action, desiccated the tracheal mucosa, decreased or brought to a standstill the flow of mucus, rendered the tracheal surface extremely vulnerable to mechanical trauma, and caused spasms of the smooth muscles of the trachea. Negative ions reportedly had opposite effects and counteracted the action of positive air ionization. The authors also observed that positive ions increase the level of the potent neurohormone 5-hydroxytryptamine (5-HT), or serotonin, in the exposed tracheal tissue and that 5-HT is the main agent causing the functional changes produced by positively charged air ions. Negative ions were reported to accelerate reoxidation of reduced cytochrome c and the cytochrome-mediated conversion of succinate to fumarate, by direct action on cytochrome c oxidase. Negative ions were also observed to accelerate the catabolism of free 5-HT, resulting in increased urinary excretion of its specific metabolite 5-hydroxyindoleacetic acid. Krueger proposed that the accelerated oxidation of 5-HT is due to an interaction of negative ions with cytochrome c oxidase.

Krueger's group also reported significant growth stimulation in cultivated oat (*Avena sativa*), barley (*Hordeum vulgare*), lettuce (*Latuca sativa*) and pea (*Pisum sativum*) through interaction of air ions of either polarity with cytochrome c and other cell enzymes. According to Krueger et al., development of iron chlorosis in *H. vulgare,* growing in an iron-free nutrient medium was markedly accelerated under air ionization with simultaneous significant increase in cytochrome c content. Air ions of either polarity reportedly also affected the ratio of active and residual iron in these plants. When grown in an air-ion-depleted environment, *H. vulgare* showed significant growth retardation and lack of turgor. Larval growth, biosynthesis of catalase, peroxidase, and cytochrome c oxidase and onset of spinning in the silk worm (*Bombyx mori*) were reported to be significantly and reproduciby accelerated under air ionization. These findings support earlier experimental evidence suggesting that the important enzyme of cellular respiration, cytochrome c, is a chief mediator for the biological action of air ions. Results of Krueger's experiments with chloroplasts suggest a stimulation of adenosine triophosphate metabolism by air ions. The implications of these findings, if confirmed by further experiments and found to be valid for other cell systems, would be quite apparent.

Relaxation of tracheobronchial spasm, reduced susceptibility of the tracheal mucosa to trauma, increased production of watery mucus in the respiratory tract, and stimulation of ciliary activity, as reported by Krueger et al. for negative ions, would have definite clinical implications. The same holds true for his later work in which he investigated the effects of air ions on bacterial and viral pneumonia, influenza, and coccidioidomycosis in the mouse. However, Krueger always carefully refrained from making any therapeutic claims for air ions.

Krueger's pioneering air ion research gained him international recognition and honors, among them the William F. Petersen Award of the International Society of Biometerology in 1966. However, his work did not remain without controversy. Several investigators who attempted to reproduce some of his results were unable to do so. For example, Andersen at the University of Aarhus in Denmark,[196] and Kensler and Battista in this country,[197] tried unsuccessfully to duplicate Krueger's most frequently quoted findings, namely the effects of air ions on the mucociliary function of the trachea. Wehner (unpublished data) failed to observe effects of air ions on the growth of barley (*H. vulgare*), using fresh weight, dry weight, and stem length as criteria. Sulman et al. in Israel,[198-203] and Olivereau in France,[204] on the other hand, independently confirmed the effects of air ions on serotonin levels, one of Krueger's most significant observations. Sulman and his colleagues also confirmed a significant attenuating effect of negative air ions on bacterial growth,[205] first reported in this country by Krueger et al. in 1957.[143] Wehner et al.[206] observed improved pulmonary clearance of $^{239}PuO_2$ in rats exposed to negatively charged aerosol, results which would tend to support Krueger's findings of improved mucociliary function.

Igho Hart Kornblueh (1902—1973) left his mark on a different area of air ion research. He became internationally known for his advocacy of air ion therapy in clinical medicine. Born in Austria, he graduated from the University of Vienna School of Medicine in 1923 and immigrated to the United States in 1935. He was Medical Director of the Department of Physical Medicine and Rehabilitation at the Graduate Hospital, University of Pennsylvania, from 1953 to 1967, and consultant to that Department from 1967 until his death. He was also Chief of the Department of Physical Medicine and Rehabilitation at the Northeastern Hospital in Philadelphia.

Due to his European background, Kornblueh grew up in a medical tradition which considers areas of physical medicine such as climatotherapy, balneology, and kurortology as integral parts of the physician's therapeutic armamentarium. He believed that the strong skepticism of his American colleagues toward these fields deprived their patients of therapeutic modalities which are generally recognized in Europe as safe and efficacious when prudently applied by experienced physicians. "Al, everything that goes beyond an injection, a pill, or the scalpel is considered quackery here," he once told me with a touch of exasperation.

It was not Kornblueh's nature to lament from the sidelines. Instead, he energetically embarked on a crusade for a better appreciation of the potential of physical therapy in general and air ion therapy in particular. He was one of the founders of — and until his death the driving force behind — the American Institute of Medical Climatology in Philadelphia. He initiated and organized the International Conference on Ionization of the Air in 1961, the first (and to this date the only) international meeting in this country exclusively devoted to air ion research. Over the years, he never missed an opportunity to call attention to the cause in which he so strongly believed.

Kornblueh was not the first one in this country to advocate air ion therapy and only a few original clinical research publications originated from his pen. Yet, his paper on the relief of hayfever symptoms in negatively ionized rooms[207] has become the most frequently referenced clinical publications in the field. He and his co-workers also pioneered in the therapeutic use of air ions in burned patients[208,209] and in postoperative discomfort,[210,211] and described effects of air ions on the encephalogram.[212] The influence of air ions on the encephalogram was subsequently confirmed by Assael et al.[213] in Israel.

Of course, there have been numerous investigators engaged in air ion research in a number of countries. An overview of their work, described in articles and books elsewhere,[2,214-216] would exceed the scope and intentions of this chronicle. While some of these investigators have made significant contributions to air ion research (e.g., Tyczka, a biometeorologist in Poland, and Deleanu, a clinician in Romania), none of them matched the stature and international recognition achieved by Krueger and Kornblueh. With their demise an era had come to an end.

V. HYDROAEROIONS/ELECTRO-AEROSOLS

A chronicle of air ion research would be incomplete without mentioning the emergence of the hydroaeroions, more commonly referred to as electro-aerosols. Electro-aerosol therapy[2,217,218] (EAT) evolved in Germany from air ion therapy in the late 1930s and 1940s. Its rationale was based on the concept of combining the beneficial physiological action of air ions with the well-known and well-documented therapeutic effects of conventional aerosol therapy (croup tent, humidifier, vaporizer, etc.) EAT comprises the inhalation exposure of patients to aqueous droplets of respirable size which carry high unipolar electric charges, usually of negative polarity. Mineral waters from springs in spas usually serve as the liquid phase in European EAT clinics; Soviet clinicians have been using mostly plain or distilled water, and in the U.S. we generally used a slightly hypertonic sodium chloride solution.

Indications of EAT include bronchial asthma, various forms of bronchitis, croup, whooping cough, emphysema, upper respiratory infections such as the common cold, influenza, and sinusitis, pneumoconioses, bronchiectasis, empyema, cyctic fibrosis and allergic conditions of the respiratory tract such as hayfever and postnasal drip. In Europe, indications for EAT also include various forms of autonomic dysfunction. The autonomic system regulates involuntary physiological functions such as blood pressure, body temperature, and intestinal activity. Malfunctions in this complex central regulating mechanism can result in peripheral consequences ranging from sleeplessness, nervous tension and abnormal blood pressure to constipation and certain (autonomically, not organically, caused) menstrual disorders. Experimental and clinical evidence suggests that EAT acts not only locally on the target tissue in the respiratory tract but also centrally via the autonomic system.[217,219-221]

Results in hundreds of thousands of EAT patients, observed independently by clinicians in several countries over a period of more than 4 decades, demonstrate the significantly greater efficacy of EAT over air ion therapy. For example, Kornblueh's patients experienced relief from their hayfever symptoms for only as long as they remained in the ionized treatment room.[207] EAT patients, by comparison, generally report up to 6 months relief, following a series of 15 to 20 45-min treatments.[1] Furthermore, while therapeutic effects of air ions often are marginal at best, EAT generally achieves readily apparent and lasting results. The following differences between these basically related modalities may explain their different efficacies:

1. The concentration of biologically active so-called "small" air ions is inversely proportional to the ambient relative humidity and the number of condensation nuclei. For this reason, and for reasons outlined below, air ion researchers have to take intricate precautions to demonstrate biological effects of air ions. Freedom from ambient environmental interferences (relative humidity, air pollution, air currents, electric fields) is usually not achievable in a clinic.

2. Small and intermediate-size air ions, with diameters ranging from 0.001 to 0.03 μm, have very little mass. They are, therefore, extremely susceptible to even the slightest air currents (air conditioning and heating systems, drafts, convection currents) and environmental electric fields. Electro-aerosols, by comparison, have a mass median diameter of 3 to 4 μm.[223] In addition, they are, in the most commonly used electro-aerosol generating systems, propelled toward their target by an air pressure of 45 psi. This gives the droplets additional directional stability. They are, therefore, much more independent of, and resistent to, environmental conditions and interferences, than air ions.
3. The aqueous droplets of electro-aerosols are physiologically more advantageous vehicles for the electric charge than the submicroscopic particles or gas clusters of air ions. This statement is based on the observation that the inhalation of aerosols per se — that is, without intentionally introduced electric charge — is a widely used, efficacious therapy. EAT thus combines two physiologically active agents, the electric charge and the aerosol proper. In addition, the choice of an appropriate liquid phase (water, hypotonic, isotonic or hypertonic saline solutions for nonspecific therapy, medicated solutions for specific therapy) facilitates a therapeutic flexibility not afforded by air ions.
4. A typical air ion carries between one and six electronic charges.[224,225] A typical electro-aerosol droplet from a Küster-Barthel generator carries several hundred electronic charges. Electro-aerosols thus deposit a much higher "dose" of electric charges at the target tissue than air ions.

Developed by Küster, a medical professor at the Goethe University in Frankfurt, in collaboration with Hans Barthel, an inventor/engineer, and clinically pioneered by Himsted,[219] Schultz,[220] and Bisa,[221,226] EAT reached its zenith in Europe in the 1950s. Several spas (e.g., Wiesbaden and Bad Ems in West Germany) installed so-called inhalatoria in which up to 25 patients can be treated simultaneously. Several of the larger mining companies in West Germany's industrial Ruhr area had installed similar facilities under the direction of the companies' medical directors for the prevention and treatment of silicosis in their miners.[227,228] Private practitioners — most prominently among them Schultz and Bisa — and hospitals are also using EAT which is recognized and reimbursed by the German insurance companies.

By the 1960s, it had become relatively quiet about EAT in Europe. My interviews in 1964 with several medical directors of mining companies and public inhalatoria as well as with the most prominent practitioners of EAT revealed (1) EAT in spas and in private clinics continued to yield gratifying results; however, the physicians saw no need to repeat in continued publications what already had been documented and published repeatedly in the past; (2) a different situation had developed in the mining industry. Turf fights, political pressures, and arguments between management and the labor union about whether or not the miners should receive pay for time spent in the inhalatoria, had all but shut down the EAT inhalatoria in the mining companies.

In 1961 I introduced EAT to the U.S. In the following 7 years approximately 1000 patients received EAT in the greater Dallas area.[1,229] While it was gratifying to see consistently good (and sometimes dramatic) results in so many patients — most of whom were referred to us because they failed to respond to conventional therapy — I failed in my most important objective. This objective was to generate enough interest to induce a medical school or research hospital to investigate the efficacy of EAT in a controlled double-blind study which we were unable to conduct, given the limitations and different purpose of our facilities. Controlled double-blind studies are the acid test which any therapy has to pass before its acceptance by the medical establishment. At one time this objective seemed within grasp when the Veterans Administration Hospital in McKinney, Texas, embarked on such a study. However, in 1964 during the first phase of the project, which showed good preliminary results,[230,231] the hospital was closed because of nationwide economy measures. With my departure from Texas in 1967 EAT in this country fell dormant.

VI. THE PRESENT AND THE LESSONS OF HISTORY

Where, then, do we stand today and what are the lessons of history?

Today, a new generation of scientists has taken over. They are competent in their specialties and are aided in their research by state-of-the-art electronics and other equipment. Therefore, the prerequisites for good research exist. However, the question of continuity of efforts arises. Krueger, for example, provided continuity of effort which develops the precious commodity called experience. Based on the level of continued research activity, as reflected in the number of technical publications, the center of gravity in air ion research now appears to have shifted to the Soviet Union. The group of Reinet, Sibul, and their colleagues at Tartu State University in Estonia has been particularly active in air ion and EAT research. Through personal contact over a number of years I became familiar with some of their work. Unfortunately, only a small fraction of it is accessible to us in English, and then only in abstracts.

As to the lessons of history, they are pretty much in the eyes of the beholder. Hundreds of publications over 6 decades of air ion research invite the conclusion that "there must be something to it." Sheer numbers are, of course, an unacceptable substitute for quality and scientific proof, and there are several groups of investigators who reported that they were unable to observe ion effects in their experiments,[3,196,197,232-237] not to mention the unknown number of those who chose not to report lack of effects. Investigations of indoor air quality during the last 10 years have demonstrated numerous pollutants contaminating indoor air.[238-242] These pollutants include radon and radon daughters, and a multitude of combustion products including tobacco smoke, and many biogenic particles and organics. Many of these compounds are biologically active. The interaction of air ions with air pollutants needs to be investigated. Numerous toxicants, carcinogens and pathogens at uncontrolled and widely varying (albeit low, but chronic) concentrations might interact with air ions. It would be rather difficult to identify the causative agent of any biological effects observed under these circumstances.

On the other hand, it hardly appears justified to summarily dismiss all air ion research as poor and/or inconclusive. Furthermore, several independently conducted, technically less-than-perfect studies might, under certain circumstances, still reveal a valid trend or the existence of a phenomenon. Some of the opponents of this field theorize that the laws of physics preclude any biological effects of air ions. Krueger used to counter this by the argument that, according to the laws of aerodynamics, a bumblebee would be unable to fly. The laws of physics, of course, never fail us, only our interpretation of them. Others maintain that typical concentrations of air ions in the atmosphere (1 ion:2.7×10^{13} molecules of air) are too low to have biological effects. This contention can be rebuffed in principle by reference to the male silk worm (*Bombyx mori*) which reacts to the sex attractant of the female at concentrations of one molecule in 1.4×10^{17} molecules of air.[239] A more appropriate question concerns the practical significance of air ion effects. If air ions are so fickle, and if their biological effects are so subtle that they are so difficult to demonstrate to the satisfaction of the scientific community, will they ever amount to more than a scientific curiosity of merely academic interest?

History as recent as the 10th International Congress of Biometeorology, July 26 to 30, 1984, in Tokyo has shown that the controversy surrounding air ion research is as lively today as it was 20, 30, or 40 years ago. Anyone who has been associated with this field for 35 years, as I have been in one form or another, is bound to form certain opinions. I have tried not to let my opinions get the better of my pen. This would be editorializing and I do not think that history should be editorialized. One thing appears certain, though. The scientists who now represent the cutting edge of technical and biomedical developments in air ion research, continue to face great challenges. They need and deserve all the support they can get.

REFERENCES

1. **Wehner, A. P.,** Weitere Erfahrungen mit der Elektroaerosol-Inhalationstherapie bei Erkrankungen des Atmungssystems, *Z. Biol. Aerosol Forsch.*, 13, 557, 1967.
2. **Wehner, A. P.,** Special review: electro-aerosols, air ions and physical medicine, *Am. J. Phys. Med.*, 48, 119, 1969.
3. **Wehner, A. P., Ragan, H. A., Jaffe, R. A., Weigel, R. J., and Lundstrom, D. L.,** Effects of negatively charged aerosol on blood and cerebrospinal fluid parameters in rats, *Int. J. Biometeorol.*, 27, 259, 1983.
4. **Franklin, B.,** *Philos. Trans. R. Soc.*, 47, 289, 1751.
5. **Dalibard, T. F.,** Letter to Acad. de Sci., 1752.
6. **Bertholon, Abbé, P.,** *De l'Electricite du Corps Humain dans l'Etat de Sante et de Maladie,* Bernuset, Lyon, 1780.
7. **Coulomb, C. A.,** *Mém. de l'Acad Paris, 616,* 1795.
8. **Elster, J. and Geitel, H.,** Über die Existenz elektrischer Ionen in der Atmosphäre, *Phys. Z.*, 1, 245, 1899.
9. **Czermak, P.,** Über Elektrizitätszerstreuung bei Föhn, *Mitt. Wien. Akad. Wiss.*, 27, 1901; *Phys. Z.*, 3, 185, 1902.
10. **Sokolof, A. P.,** Die Ionisation u. Radiumaktivität der atmosphärischen Luft, *Sanitätsmitteilung der russischen balneologischen Gesellschaft in Pjatigorstk,* 1904.
11. **Steffens, P.,** *Witterungswechsel und Rheumatismus,* Leipzig, 1910.
12. **Dessauer, F.,** *Zehn Jahre Forschung auf dem Physikalisch-Medizinischen Grenzgebiet,* Georg Thieme Verlag, Leipzig, 1931.
13. **Tchijevsky, A. L.,** L'aéroionothérapie des maladies pulmonaires; Recherches expérimentales de l'effect de l'air ionisé sur la tuberculose des poumons, la pneumonie et la bronchite, *Presse Therm. Clim.*, 11, 121, 1929.
14. **Tchijevsky, A. L.,** Experimentelle Untersuchungen über den Einfluss ionisierter Luft auf die Lungentuberkulose, *Z. Gesamte Phys. Ther.*, 39, 46, 1930.
15. **Tchijevsky, A. L. and Utz, N.,** Observations sur des malades pulmonaires traités á l'air ionisé, *Prog. Méd.*, 33, 1439, 1931.
16. **Tchijevsky, A. L.,** Theoretical foundations of the operation of an electroeffluvial ionizer, in *Problemy ionifikatsii, I, Ts NILI,* Voronez, 1933.
17. **Tchijevsky, A. L.,** Die Wege des Eindringens von Luftionen in den Organismus und die physiologische Wirkung von Luftionen, *Acta Med. Scand.*, 83, 219, 1934.
18. **Tchijevsky, A. L. and Vognar, L.,** Le viellissement de l'organisme retardé moyennant l'inspiration d'air ionisé négativement, *Prog. Méd.*, 36, 227, 1934.
19. **Tchijevsky, A. L.,** Action de l'ionisation de l'atmosphére et de l'ionisation artificielle de l'air sur les organismes sains et les organismes malades, Masson et Cié, Paris, 1934.
20. **Tchijevsky, A. L. and Bobrov,** Influence d'un flux dirigé d'aéroions négatifs sur la vitesse de la régénération des plaies ouvertes. Théorie mathematique des processus, *Acta Med. Scand.*, 85, 452, 1935.
21. **Tchijevsky, A. L.,** Traitement des hypertensions sanguines de différente origine á l'aéroionisation de polarité négative, *Acta Med. Scand.*, 99, 117, 1939.
22. **Tchijevsky, A. L.,** *Air Ionization, its Role in the National Economy,* Gosplomizdat, Moscow, 1960.
23. **Tchijevsky, A. L., Ed.,** *Transactions of the Central Laboratory for Scientific Research on Ionification,* Vol. 1, 2, and 3, Publishing House, The Commune, Voronez, 1963.
24. **Yaglou, C. P., Benjamin, L. C., and Brandt, A.,** Observations on a group of subjects before, during and after exposure to ionized air, *J. Ind. Hyg.*, 15, 8, 1933.
25. **Edstrom, G.,** Der Einfluss unipolar beladener Luft auf die Chronaxie motorischer Nerven, *Z. Gesamte Phys. Ther.*, 42, 124, 1932.
26. **Edstrom, G.,** Studies in natural and artificial atmospheric electric ions, *Acta Med. Scand.*, Suppl. 61, 1935.

27. **Edstrom, G.**, Investigations into the effect of unipolarly charged air on the surface temperature, *Acta Med. Scand.*, 83, 523, 1934.
28. **Leicher, H.**, Über Inhalation von negativ ionisierter Luft (Anionenbehandlung) bei Nasen und Ohrenkrankheiten, *Z. Laryngol.*, 22, 497, 1932.
29. **Strasburger, and Lampert, H.**, Weitere klinisch-therapeutische Erfahrungen bei der Einatmung negativ ionisierter Luft, *Dtsch. Med. Z.*, 59, 1316, 1933.
30. **Wilhelmi, A. and Wilhelmi, E.**, Therapeutische Erfahrungen mit unipolar ionisierter Luft, *Fortschr. Ther.*, 9, 528, 1933.
31. **Fervers, C.**, Experimentelle Unterschuchungen über psycho-phsyische Beeinflussung durch ionisierte Luft, *Dtsch. Med. Z.*, 1876, 1933.
32. **Lampert, H.**, *Physikalische Therapie*, Steinkopff Verlag, Dresden, 1938.
33. **Landsmann, E.**, Über die Behandlung von Bronchialasthma mit Aeroionisation, *Wien. Klin. Z.*, 48, 1348, 1935.
34. **Achler, R.**, Über die Wirkung der Einatmung Unipolar Gleandener Luft auf die Durchblutungsgrösse der Schleimaute des Atemtraktus, Dissertation, Marburg, 1940.
35. **Baumgarten, F.**, Untersuchungen über die Einwirkung Unipolar Geladener Luft auf die Resistenz mit Bacterium Enteritidis Gärtner Infizierter Mäuse, Dissertation, Marburg, 1940.
35a. **Vilenkin, L. and Rossiisky, D.**, Electro-aeroionotherapy for influenza, *Acta Med. Scand.*, 89, 190, 1936.
36. **Kaufmann, K., Nekludow, B., Nekludowa, P. O., and Boschwolow, A. S.**, Unipolar geladene Luft als die Methode der Tuberkulose Behandlung, die Veränderung des Zahlenverhältnisses der weissen Blutkörperchen unter dem Einfluss der unipolar geladenen Luft und der Zerstäubung von Ca-Präparaten, *Balneologe*, 2, 139, 1935.
36a. **Nekludow, W. N., Nekludowa, E. A., and Boschewolow, A. I.**, Der Einfluss unipolar geladener Luft auf das weisse Blutbild bei Meerschweinchen, *Virchows Arch. Pathol. Anat. Physiol.*, 292, 438, 1934.
37. **Pachinkov, E. I.**, Traitement de l'asthme bronchique par l'aéroionisation, *Arch. Thér.*, 12, 140, 1934.
38. **Prosorovsky, B. M.**, Traitement des troubles végétatifs endocriniens et de l'asthme bronchique par l'aéroionisation, *Probl. d'ionisation, Voronége*, 8, 26, 1934.
39. **Oserkowsky, J.**, Quantitative relation between chlorophyll and iron in green and chlorotic pear leaves, *Plant Physiol.*, 8, 449, 1933.
40. **Sapozhnikova, K. V.**, Respiration of wheat seed in ionized air, *C. R. Acad. Sci. U.S.S.R.*, 8, 119, 1928.
41. **Vysotskii, K. A.**, The electro-ionization of the cotton plant, *Proc. Turkestan Plant Breeding Station Central Cotton Committee, U.S.S.R.*, 13, 109, 1928.
42. **Vysotskii, K. A.**, Electro-ionization of cotton, *Acta Univ. Asiae Med.*, (Ser. 10), 1, 3, 1929.
43. **Zvonitzky, N. S. and Obrossov, A. N.**, Hochionisierte Luft als biologischer Faktor, *Z. Phys. Diatet. Ther.*, 44, 176, 1933.
44. **Abe, S.**, Inhibitory effects of air ions on tumor development. I, *Ikai Jiho*, Nov., 2150, 1935.
45. **Abe, S.**, Inhibitory effects of air ions on tumor development. II, *Nippon Igaku Shimpoh*, Feb., 701, 1936.
46. **Abe, S.**, Air ion effects on blood sugar and growth of adult guinea pigs, *Hokkaido Igaku Zasshi*, 10, 2154, 1936.
47. **Anazawa, Y.**, Hypotension and air ion bathing, *Chiryogaku-Zasshi*, 7, 611, 1936.
48. **Anazawa, Y.**, Report on the experience of hypertension treatment with air ions, *Rinsho Igaku*, 1, 101, 1938.
49. **Anazawa, Y.**, Constipation and air ions, *Shinryo*, 11, 250, 1936.
50. **Ariizumi, K.**, Air ion effects on enzyme solution, especially on amylase, *Hokkaido Igaku Zasshi*, 11, 2367, 1936.
51. **Ashiba, M.**, Blood pressure-decreasing effect of air ions and its quantitative relation, *Ikai Jiho*, 121, 2214, 1937.
52. **Ashiba, M.**, Studies on air ion effects, *Hokkaido Igaku Zasshi*, 9, 2344, 1937.
53. **Ashiba, M.**, Air ion effects on the frog heart, especially the excised frog heart, *Hokkaido Igaku Zasshi*, 12, 3456, 1937.
54. **Fukui, T.**, Physiological effects of air ions. I. Effects on respiration of normal rabbits, *Hokkaido Igaku Zasshi*, 3, 662, 1937.
55. **Fukui, T.**, Physiological effects of air ions. II. Air ion effects on respiration of animals dosed with drugs, *Hokkaido Igaku Zasshi*, 9, 2280, 1937.
56. **Fukui, T.**, Physiological effects of air ions. III. Diuretic effect, *Hokkaido Igaku Zasshi*, 11, 3180, 1937.
57. **Fukui, T. and Matsushima, I.**, Air ion effect on rabbit urination and ear blood vessels, *Nippon Rengoeiseigakkaishi*, 9, 12, 1937.
58. **Fukuzumi, I.**, Effects of ionized air on bacterial growth, *Hokkaido Igaku Zasshi*, 6, 1381, 1936.
59. **Fukuzumi, I.**, Air ion effects on the function of the reticuloendothelial system, *Hokkaido Igaku Zasshi*, 10, 1603, 1937.
60. **Fukuzumi, I.**, Air ion effects on experimental anemia, *Hokkaido Igaku Zasshi*, 1, 21, 1938.

61. **Honma, S.,** Air ion effect on blood sugar, *Hokkaido Igaku Zasshi,* 7, 1684, 1937.
62. **Honma, S. and Kitahara, T.,** Air ion effects on biological bodies, *Nippon Naika Gakkai Zasshi,* 24, 415, 1936.
63. **Honma, S., Kitahara, T., and Aikawa, O.,** Air ion effects on biological bodies (effects on blood pressure, pulse, etc.), *Hokkaido Igaku Zasshi,* 11, 2303, 1936.
64. **Horimi, T. and Hasegawa, T.,** Experience of air ion treatment of hypertension. I, *Seishin Shinkeigaku Zasshi,* 41, 249, 1937.
65. **Horimi, T. and Hasegawa, T.,** Experiences of air ion treatment of hypertension. II, *Rinsho Nippon Igaku,* 6, 168, 1937.
66. **Miyoshi, K.,** Experience of air ion inhalation therapy of hypertension, *Chiryogaku Zasshi,* 6, 1377, 1936.
67. **Nakajima, Y.,** Hypertension treatment with negative air ions, *Nippon Naika Gakkai Zasshi,* 22, 164, 1934.
68. **Nakajima, Y., Kaga, Y., and Irie, H.,** Negative air ion treatment of hypertension patients, *Nippon Naika Gakkai Zasshi,* 20, 260, 1932.
69. **Ogawa, S.,** Hypertension and air ion therapy, *Shinryo-to-Keiken,* 1, 476, 1937.
70. **Kusaka, S. and Itano, S.,** Negative air ion treatment of hypertension, *Okayama Igakkai Zasshi,* 2, 461, 1937.
71. **Ando, M.,** Hypertension treatment by air ions, *Rinsho Igaku,* 11, 2367, 1936.
72. **Ishidate, F.,** Air ion effects on complement and blood composition in the guinea pig, *Ikai-Hiho,* 22, 28, 1937.
73. **Ishidate, F.,** Air ion effects on guinea pig blood composition with special reference to complements, *Ikai-Jiho,* 2229, 1937.
74. **Karube, S.,** Air ion effects on blood composition in rabbits, *Nippon Rengo-Eiseigakkai-Shi,* 9, 76, 1937.
75. **Karube, S.,** Air ion effects on blood composition in rabbits, *Kobe-shiritsu Eisei Shikenjo Hokoku,* 1, 65, 1937.
76. **Murayama, T.,** Air ion effects on the development of animal skeletal structure, *Hokkaido Igaku Zasshi,* 6, 1381, 1936.
77. **Nakajima, Y.,** Air ion effect on acid-base equilibrium of blood, *Hokkaido Igaku Zasshi,* 6, 1317, 1935.
78. **Sakuma, T.,** Air ion effects on physicochemical properties of blood, especially on blood coagulation. I. Blood coagulation time, *Hokkaido Igaku Zasshi,* 5, 1237, 1937.
79. **Sakuma, T.,** Air ion effects on physicochemical properties of blood, especially on blood coagulation. II. Blood coagulation elements, *Hokkaido Igaku Zasshi,* 5, 1254, 1937.
80. **Sakuma, T.,** Air ion effects on physicohemical characteristics of blood, especially on blood coagulation. III. Platelets as blood coagulation elements, *Hokkaido Igaku Zasshi,* 8, 2051, 1937.
81. **Sakuma, T.,** Air ion effects on physicochemical properties of blood, especially on serum surface tension, *Hokkaido Igaku Zasshi,* 5, 1274, 1937.
82. **Sakuma, T.,** Air ion effect on physicochemical properties of blood, especially refractive index and the total protein content of serum, *Hokkaido Igaku Zasshi,* 8, 2058, 1937.
83. **Sakuma, T.,** Air ion effects on physicochemical properties of blood, especially on serum viscosity, *Hokkaido Igaku Zasshi,* 8, 2071, 1937.
84. **Sekino, H. and Nakamura, S.,** Negative air ion effects on erythrocyte sedimentation time in surgical tuberculosis patients, *Ikai-Jiho,* August, 2191, 1936.
85. **Suzuki, T.,** Air ion effects on the glycogen level, *Hokkaido Igaku Zasshi,* 4, 1021, 1936.
86. **Takahashi, K.,** Air ion effects on rabbit blood components. I. Air ion effects on potassium and calcium levels in serum, *Hokkaido Igaku Zasshi,* 6, 128, 1936.
87. **Takahashi, K.,** Air ion effects on rabbit blood components. II. Air ion effects on iodic acid value of rabbit blood, *Hokkaido Igaku Zasshi,* 9, 2020, 1936.
88. **Takahashi, K.,** Air ion effects on rabbit blood components. III. Air ion effects on the lactic acid content of rabbit blood, *Hokkaido Igaku Zasshi,* 9, 2031, 1936.
89. **Takasaki, O.,** Negative air ion effects on blood regeneration, *Nippon Rentogen-gakkai Zasshi,* 15, 218, 1937.
90. **Tsuchida, R.,** Air ion effects on enzymes of rabbit blood, *Hokkaido Igaku Zasshi,* 12, 2767, 1937.
91. **Matsugaki, T. and Hasuo, H.,** Air ion effects on spasmic coughing, *Jicchi Ika-to-Rinshyo,* 13, 1098, 1936.
92. **Kishioka, K.,** Value of air ion therapy, *Gunidan Zasshi,* 285, 223, 1937.
93. **Kojima, H.,** Pneumonia and air ion treatment, *Shinryo,* 9, 307, 1937.
94. **Matsuhashi, T.,** Air ion effects on immunoserum value and Pneumococcus infection, *Nippon Rengo-Eiseigakkai-Shi,* 8, 24, 1936.
95. **Matsui, S.,** Negative air ion effects on the growth of weak juvenile children, *Ikai-Jiho,* June, 2234, 1937.
96. **Minamide, H. and Isoya, T.,** Clinical application of air ions in pediatrics, *Tohoku Igaku Zasshi,* 18 (Suppl.), 133, 1935.
97. **Minamiura, K.,** Clinical application of artificially charged aerosol, *Hokkaido Igaku Zasshi,* 14, 2309, 1935.

98. **Minamiura, K.,** Therapeutic application of air ions, *Chiryo-oyobi-Shoho,* 184, 1038, 1935.
99. **Miyoshi, K.,** Precautions for air ion therapy, *Rinsho-to-Yakubutsu,* 8, 635, 1936.
100. **Miyoshi, K.,** What is the air ion therapy? Diseases curable by air ion therapy, *Kansai Iji,* 295, 3, 1936.
101. **Miyoshi, K.,** Experimental and clinical studies on air ion effects on biological bodies, *Nippon Iji Shinpo,* 743, 4201, 1936.
102. **Mori, A.,** Air ion therapy, *Chiryogaku Zasshi,* 1, 25, 1935.
103. **Murakoshi, Y.,** Clinical and experimental studies on the effects of air ion inhalation on the function of vegetative nerve system and blood components, *Tokyo Iji Shinshi,* 29, 2029, 1936.
104. **Ohnishi, S.,** Therapeutic effects of air ion, *Chiryogaku Zasshi,* August, 236, 1935.
105. **Ohta, M.,** Weather diseases, air electricity, and ion therapy, *Chiryo-oyobi-Shoho,* 199, 1824, 1936.
106. **Takahashi, K.,** Air ion effect on allergy, *Hokkaido Igaku Zasshi,* 9, 2041, 1936.
107. **Togasawa, K.,** Negative ion inhalation equipment and experimental and clinical experiments, *Naikagakkai Zasshi,* 3, 414, 1935.
108. **Togasawa, K.,** Clinical case report on air ion treatment, *Iji-Shinpo,* 670, 2003, 1935.
109. **Kimura, S. and Taniguchi, M.,** Theory and practice of air ion therapy in *Nanzendo Shoten,* Tokyo, 1938, 257.
110. **Kakita, H.,** Clinical reports on Togasawa's ionotherapy, *Kaigus-Gunikai Zasshi,* 25, 628, 1936.
111. **Jo, K.,** Air ion effects on luminescence bacteria, *Nippon Eiseigaku Zasshi,* 9, 124, 1937.
112. **Kimura, S.,** Effects of atmospheric electricity on living bodies, *Nippon Onsen Kikogakkai Zasshi,* 2, 108, 1936.
113. **Kimura, S., Ashiba, M., and Matsushima, J.,** Influence of the air lacking light ions and the effect of its artificial ionization upon human beings in occupied rooms, *Jpn. J. Med. Sci. Biol.,* 3(7), 1, 1939.
114. **Ohnishi, S., Tsukada, J., and Igarashi, Y.,** City hygienic studies of air ions, *Nippon Rengo-Eiseigakkai Zasshi,* 8, 117, 1936.
115. **Ohnishi, S., Tsukada, J., and Igarashi, Y.,** Effects of heating systems on room air ion counts, *Nippon Rengo-Eiseigakkai-Shi,* 9, 124, 1937.
116. **Ohnishi, S., Tsukada, J., and Igarashi, Y.,** City hygienic studies. II. Changes in air ion count in Tokyo with time and seasons, *Nippon Rengo-Eiseigakkai-Shi,* 9, 124, 1937.
117. **Moritani, H.,** Air ion effects on bacterial growth. II. Air ion effects on Escherichia coli, *Hokkaido Igaku Zasshi,* 5, 1295, 1937.
118. **Moritani, H.,** Air ions and experimental ricket, *Hokkaido Igaku Zasshi,* 8, 1773, 1936.
119. **Moritani, H.,** Air ions and experimental ricket. II. Air ion effects on ergosterin activity, *Hokkaido Igaku Zasshi,* 5, 1288, 1937.
120. **Moritani, H.,** Air ions and experimental ricket. I. Effect of air ion on experimental ricket induction, *Shohnika Zasshi,* 43, 20, 1937.
121. **Ohmura, K.,** Air ion effects on protein metabolism of rabbits. I. Air ion effects on nitrogen metabolism of rabbits, *Hokkaido Igaku Zasshi,* 7, 1822, 1937.
122. **Ohmura, K.,** Air ion effects on protein metabolism of rabbits. II. Air ion effects on creatinin metabolism, *Hokkaido Igaku Zasshi,* 7, 1845, 1937.
123. **Ohmura, K.,** Air ion effects on protein metabolism of rabbits. III. Air ion effects on the fate of urinary urea in rabbits, *Hokkaido Igaku Zasshi,* 7, 1880, 1937.
124. **Murayama, T.,** Air ion effects on total cholesterin and fatty acid in blood, *Hokkaido Igaku Zasshi,* 8, 2156, 1937.
125. **Murayama, T.,** Air ion effects on the growth of guinea pig, *Hokkaido Igaku Zasshi,* 11, 3159, 1937.
126. **Ohta, O.,** Air ion effects on development and metamorphosis of toads, *Nippon Rengo-Eiseigakkai-Shi,* 9, 5, 1937.
127. **Saito, T.,** Air ion effect on chronaxy. I. Normal pigeon chronaxy, *Hokkaido Igaku Zasshi,* 8, 1968, 1937.
128. **Saito, T.,** Air ion effect on chronaxy. II. Normal pigeon chronaxy, *Hokkaida Igaku Zasshi,* 9, 2191, 1937.
129. **Saito, T.,** Air ion effect on pigeon beriberi and chronaxy, *Hokkaido Igaku Zasshi,* 10, 2469, 1937.
130. **Saito, T.,** Air ion effect on chronaxy of exhausted pigeon, *Hokkaido Igaku Zasshi,* 11, 3063, 1937.
131. **Sekino, H.,** Effect of ionized air bathing on biological bodies. I. Effect on phagocytosis of reticuloendothelial system of rabbits, *Hokkaido Igaku Zasshi,* 10, 2712, 1937.
132. **Sekino, H.,** Effects of ionized air bathing on biological bodies. II. Effects of air ions on phagocytosis of the reticuloendothelial system in rabbits, *Hokkaido Igaku Zasshi,* 10, 2722, 1937.
133. **Sekino, H.,** Effects of ionized air bathing on biological bodies. III. Effects on blood pressure of rabbits, *Hokkaido Igaku Zasshi,* 16, 10, 1938.
134. **Sone, S.,** Negative air ions on experimental hypertension in rabbits, *Nippon Hoshasenigaku-Kai Zasshi,* 4, 316, 1936.
135. **Sone, S.,** Negative air ions on experimental hypertension in rabbits, *Fukuoka Ika Daigaku Zasshi,* 29, 2078, 1936.
136. **Sinaya, M. S.,** The influence of unipolar air ions on the electrophoretic mobility of erythrocytes, *Int. J. Biometeorol.,* 11, 187, 1967.

137. **Mugikura, R.,** Air ions and experimental scurvy. I, *Hokkaido Igaku Zasshi,* 5, 2776, 1937.
138. **Mugikura, R.,** Air ions and experimental scurvy. II, *Hokkaido Igaku Zasshi,* 5, 3543, 1937.
139. **Miura, Y.,** Air ion effects on silkworm, *Hokkaido Igaku Zasshi,* 12, 2535, 1936.
140. **Yuasa, K.,** Air ion effects on perfused, excised toad heart. I, *Nippon Rengo-Eiseigakai-Shi,* 8, 25, 1936.
141. **Yuasa, K.,** Air ion effects on perfused, excised toad heart. II, *Nippon Rengo-Eiseigakai-Shi,* 9, 15, 1936.
142. **Sugano, Y.,** Air ion effect on the growth of tubercle bacilli, *Hokkaido Igaku Zasshi,* 5, 15, 1937.
143. **Krueger, A. P., Smith, R. F., and Go, I. G.,** The action of air ions on bacteria, *J. Gen. Physiol.,* 41, 359, 1957.
144. **Krueger, A. P. and Smith, R. F.,** Effects of air ions on isolated rabbit trachea, *Proc. Soc. Exp. Biol. Med.,* 96, 807, 1957.
145. **Krueger, A. P. and Smith, R. F.,** Parameters of gaseous ion effects on the mammalian trachea, *J. Gen. Physiol.,* 42, 959, 1958.
146. **Krueger, A. P. and Smith, R. F.,** The effects of air ions on the living mammalian trachea, *J. Gen. Physiol.,* 42, 69, 1958.
147. **Krueger, A. P., Hicks, W. W., and Beckett, J. C.,** Effects of unipolar air ions on microorganisms and on evaporation, *J. Franklin Inst.,* 266, 9, 1958.
148. **Krueger, A. P. and Smith, R. F.,** Effects of gaseous ions on tracheal ciliary rate, *Proc. Soc. Exp. Biol. Med.,* 98, 412, 1958.
149. **Krueger, A. P., Smith, R. F., and Millar, J. W.,** Effects of air ions on trachea of primates, *Proc. Soc. Exp. Biol. Med.,* 101, 506, 1959.
150. **Krueger, A. P., Smith, R. F., Hildebrand, G. J., and Meyers, C. E.,** Further studies on gaseous ion action on trachea, *Proc. Soc. Exp. Biol. Med.,* 102, 355, 1959.
151. **Krueger, A. P. and Smith, R. F.,** An enzymatic basis for the acceleration of ciliary activity by negative air ions, *Nature (London),* 183, 1332, 1959.
152. **Krueger, A. P. and Smith, R. F.,** The physiological significance of positive and negative ionization of the atmosphere, *J. R. Inst. Public Health,* 79, 642, 1959.
153. **Krueger, A. P. and Smith, R. F.,** Parameters of gaseous ion effects on the mammalian trachea, *J. Gen. Physiol.,* 42, 959, 1959.
154. **Krueger, A. P. and Smith, R. F.,** The biological mechanisms of air ion action. I. 5-Hydroxytryptamine as the endogenous mediator of positive air ion effects on the mammalian trachea, *J. Gen. Physiol.,* 43, 533, 1960.
155. **Krueger, A. P. and Smith, R. F.,** The biological mechanisms of air ion action. II. Negative air ion effects on the concentration and metabolism of 5-hydroxytryptamine in the mammalian respiratory tract, *J. Gen. Physiol.,* 44, 269, 1960.
156. **Krueger, A. P.,** in *The Air We Breathe,* Faber and Wilson, Eds., Charles C Thomas, Springfield, Ill., 1961.
157. **Krueger, A. P.,** Further studies on the biological effects of air ions and their implications with respect to the physical aspects of air ionization, Wesix Foundation, San Francisco, 1960.
158. **Krueger, A. P., Kotaka, S., and Andriese, P. C.,** Studies on the effects of gaseous ions on plant growth. The influence of positive and negative air ions on the growth of Avena sativa, *J. Gen. Physiol.,* 45, 879, 1962.
159. **Krueger, A. P., Beckett, J. C., Andriese, P. C., and Kotaka, S.,** The construction and operation of an air purification unit for use in studies on the biological effects of gaseous ions, *J. Gen. Physiol.,* 45, 897, 1962.
160. **Krueger, A. P.,** Air ions and physiological functions, *J. Gen. Physiol.,* 2, 233, 1962.
161. **Krueger, A. P., Kotaka, S., and Andriese, P. C.,** Some observations on the physiological effects of gaseous ions, *Int. J. Biometeorol.,* 6, 33, 1962.
162. **Krueger, A. P. and Smith, R. F.,** Studies on the effects of gaseous ions on the mammalian trachea, in *Proc. 2nd Int. Bioclimatology Congr.,* New York, Pergamon Press, 1962, 498.
163. **Krueger, A. P., Kotaka, S., and Andriese, P. C.,** A study of the mechanism of air-ion-induced growth stimulation in Hordeum vulgaris, *Int. J. Biometeorol.,* 7, 17, 1963.
164. **Krueger, A. P., Hicks, W. W., and Beckett, J. C.,** The influence of air ions on certain physiological functions, in *Biometerology-Weather, Climate, and the Living Organism,* Tromp, S. W., Ed., Elsevier., New York, 1963.
165. **Krueger, A. P., Kotaka, S., and Andriese, P. C.,** Gaseous-ion-induced stimulation of cytochrome biosynthesis, *Nature (London),* 200, 707, 1963.
166. **Krueger, A. P., Andriese, P. C., and Kotaka, S.,** The biological mechanism of air ion action. The effect of CO_2^+ in inhaled air on the blood level of 5-hydroxytryptamine in mice, *Int. J. Biometeorol.,* 7, 3, 1963.
167. **Krueger, A. P., Kotaka, S., and Andriese, P. C.,** Studies of air-ion-enhanced chlorosis. I. Active and residual iron, *Int. J. Biometeorol.,* 8, 5, 1964.
168. **Krueger, A. P., Kotaka, S.,and Andriese, P. C.,** The effect of air containing O_2^-, O_2^+ CO_2^- and CO_2^+ on the growth of seedlings of *Hordeum vulgaris, Int. J. Biometeorol.,* 8, 17, 1964.

169. **Krueger, A. P., Kotaka, S., and Andriese, P. C.,** The effect of abnormally low concentrations of air ions on the growth of *Hordeum vulgaris, Int. J. Biometeorol.*, 9, 201, 1965.
170. **Krueger, A. P.,** The influence of air ions on physiological functions, in *A Survey of Human Biometerology,* No. 160, TP78, Sargent, F. and Tromp, S. W., Eds., World Meteorology Organization, 1965, 32.
171. **Krueger, A. P., Andriese, P. C., and Kotaka, S.,** The effects of inhaling non-ionized or positvely ionized air containing 2—4% CO_2 on the blood levels of 5-hydroxytryptamine in mice, *Int. J. Biometeorol.*, 10, 17, 1966.
172. **Krueger, A. P., Kotaka, S., Nishizawa, K., Kogure, Y., Takenobu, M., and Andriese, P. C.,** Air ion effects on the growth of the silkworm, *Int. J. Biometeorol.*, 10, 29, 1966.
173. **Krueger, A. P., Kotaka, S., and Andriese, P. C.,** Studies on the biological effects of gaseous ions, Special Monograph Series, Vol. 1, Biometeorological Research Centre, Leiden, The Netherlands, 1966, 1.
174. **Krueger, A. P. and Levine, H. B.,** The effect of unipolar positively ionized air on the course of coccidioidomycosis in mice, *Int. J. Biometeorol.*, 11, 279, 1967.
175. **Krueger, A. P.,** The biological effects of gaseous ions, in *Aeroionotherapy,* Gualitierotti, R., Kornblueh, I. H., and Sirtori, C., Eds., Foundation Carlo Erba, Milano, 1968.
176. **Krueger, A. P., Kotaka, S., and Andriese, P. C.,** Small air ions: their effect on blood levels of serotonin in terms of modern physical theory, *Int. J. Biometeorol.*, 12, 225, 1968.
177. **Krueger, A. P. and Kotaka, S.,** The effects of air ions on brain levels of serotonin in mice, *Int. J. Biometeorol.*, 13, 25, 1969.
178. **Krueger, A. P., Kotaka, S., and Andriese, P. C.,** Atmospheric ions and aerosols, in *An Introduction to Experimental Aerobiology,* Dimmick, R. L. and Akers, A. B., Eds., Wiley-Interscience, New York, 1969, 100.
179. **Krueger, A. P., Kotaka, S., Reed, E. J., and Turner, S.,** The effect of air ions on bacterial and viral pneumonia in mice, *Int. J. Biometeorol.*, 14, 247, 1970.
180. **Krueger, A. P. and Reed, E. J.,** Effects of the air ion environment on influenza in the mouse, *Int. J. Biometeorol.*, 16, 209, 1972.
181. **Krueger, A. P., Reed, E. J., Day, M. B., and Brook, K. A.,** Further observations on the effect of air ions on influenza in the mouse, *Int. J. Biometeorol.*, 18, 46, 1974.
182. **Krueger, A. P., Reed, E. J., Brook, K. A., and Day, M. B.,** Air ion action on bacteria, *Int. J. Biometeorol.*, 19, 65, 1975.
183. **Krueger, A. P. and Reed, E. J.,** Biological impact of small air ions, *Science,* 193, 1209, 1976.
184. **Kotaka, S., Krueger, A. P., and Andriese, P. C.,** Catalase activity of a crude enzyme preparation from iron-chlorotic barley (*Hordeum vulgare*) seedlings, *Nature (London),* 204, 1198, 1964.
185. **Kotaka, S., Krueger, A. P., and Andriese, P. C.,** Effect of air ions on IAA content of barley seedling, *Plant Cell Physiol.*, 6, 711, 1965.
186. **Kotaka, S., Krueger, A. P., and Andriese, P. C.,** Some observation on air-ion enhanced iron chlorosis in barley (*Hordeum vulgare*) seedlings, *Plant Cell Physiol.*, 6, 233, 1965.
187. **Kotaka, S., Krueger, A. P., Nisizawa, K., Ochuchi, T., Kogure, Y., Takenobu, M., and Andriese, P. C.,** Air ion effects on the oxygen consumption of barley seedlings, *Nature (London),* 208, 1112, 1965.
188. **Kotaka, S., Krueger, A. P., Nisizawa, K., Ochuchi, T., Kogure, Y., Takenobu, M., and Andriese, P. C.,** Air ion effects on the iron metabolism of barley (*Hordeum vulgare*) seedlings, *Proc. 30th Annu. Meet. Botanical Society of Japan,* Japanese Botanical Society, Tokyo, 1965, C21.
189. **Kotaka, S., Krueger, A. P., and Andriese, P. C.,** The swelling action of light on shrunken spinach chloroplasts and the effect of unipolar ionized air on this phenomenon, *Proc. 7th Int. Congr. Biochemistry,* Japanese Biochemical Society, Tokyo, 1967.
190. **Kotaka, S. and Krueger, A. P.,** Air ion effects on EDTA-induced bleaching in green barley leaves, *Int. J. Biometeorol.*, 12, 331, 1968.
191. **Kotaka, S., Krueger, A. P., and Andriese, P. C.,** The effects of air ions on light-induced swelling and dark-induced shrinking of isolated chloroplasts, *Int. J. Biometeorol.*, 12, 35, 1968.
192. **Kotaka, S., Krueger, A. P., and Andriese, P. C.,** Bleaching effect of EDTA on green barley leaves, *Plant Physiol.*, 43 (Suppl. S-4), 21, 1968.
193. **Kotaka, S., and Krueger, A. P.,** Effects of ethylenediaminetetraacetic acid (EDTA) on nucleic acid content of green barley leaves, *Proc. 11th Int. Botany Congr.*, American Society of Botanists, 1969, 115.
194. **Kotaka, S. and Krueger, A. P.,** Some observations on the bleaching effects of ethylenediaminetetraacetic acid on green barley leaves, *Plant Physiol.*, 44, 809, 1969.
195. **Kotaka, S. and Krueger, A. P.,** Air ion effects on RNase activity in green barley leaves, *Int. J. Biometeorol.*, 16, 1, 1972.
196. **Andersen, I.,** *Mucociliary Function in Trachea Exposed to Ionized and Non-Ionized Air,* Akademisk Boghandel, Aarhus, Denmark, 1971.
197. **Kensler, C. J. and Battista, S. P.,** Chemical and physical factors affecting mammalian ciliary activity, *Am. Rev. Respir. Dis.*, 93, 93, 1962.

198. **Sulman, F. G. and Pfeifer, Y.**, Ionometry of hot dry desert winds (Khamsin, Sharav) and its application to ionizing treatment of weather-sensitive patients, *Isr. J. Med. Sci.*, 9, 686, 1973.
199. **Sulman, F. G.**, Föhnleiden, ihre Ursachen und Behandlung, *Phys. Med. Rehabil.*, 12, 256, 1974.
200. **Sulman, F. G., Levy, D., Levy, A., Pfeifer, Y., Superstine, E., and Tal, E.**, Air-ionometry of hot, dry desert winds (Sharav) and treatment with air ions of weather-sensitive subjects, *Int. J. Biometeorol.*, 18, 313, 1974.
201. **Sulman, F. G.**, Climactic factors in the incidence of attacks of migraine, *Hemicrania*, 6, 2, 1974.
202. **Sulman, F. G., Levy, D., Pfeifer, Y., Superstine, E., and Tal, E.**, Effects of the Sharav and Bora on urinary neurohormone excretion in 500 weather-sensitive females, *Int. J. Biometeorol.*, 19, 202, 1975.
203. **Tal, E., Pfeifer, Y., and Sulman, F. G.**, Effect of air ionization on blood serotonin in vitro, *Experientia*, 32/3, 326, 1976.
204. **Olivereau, J. M.**, Incidence Psychophysiologiques de l'Ionisation Atmosphérique, dissertation, Universite Paris VI, 1971.
205. **Sulman, D. F., Pfeifer, Y., Shalita, B., and Tal, E.**, Air sterilization: influences of negative ionization on bacterial counts on agar plates exposed to air, *IRCS*, 2, 142, 1974.
206. **Wehner, A. P.**, Negatively charged aerosols. Effect on pulmonary clearance of inhaled $^{239}PuO_2$ in rats, *Chest*, 60, 468, 1971.
207. **Kornblueh, I. H., Piersol, G. M., and Speicher, F. P.**, Relief from pollinosis in negatively ionized rooms, *Am. J. Phys. Med.*, 37, 18, 1958.
208. **Minehart, J. R., David, T. A., and Kornblueh, I. H.**, Artificial ionization and the burned patient, *Med. Sci.*, 3, 363, 1958.
209. **David, T. A., Minehart, J. R., and Kornblueh, I. H.**, Polarized air as an adjunct in the treatment of burns, *Am. J. Phys. Med.*, 39, 111, 1960.
210. **Minehart, J. R., David, T. A., McGurk, F. C. J., and Kornblueh, I. H.**, The effect of artificially ionized air on post-operative discomfort, *Am. J. Phys. Med.*, 40, 56, 1961.
211. **David, T. A., Derau, J. V., Kornblueh, I. H., McGurk, F. C. J., and Minehart, J. R.**, Ionization of the air. The sedating effect of polarized air, in *Biometeorology*, Oxford, 1962, 484.
212. **Silverman, D. and Kornblueh, I. H.**, Effect of artificial ionization of the air on the electro-encephalogram, *Am. J. Phys. Med.*, 36, 352, 1957.
213. **Assael, M., Pfeifer, Y., and Sulman, F. G.**, Influence of artificial air ionization on the human encephalogram, *Int. J. Biometeorol.*, 18, 306, 1974.
214. **Kotaka, S.**, Effects of air ions on microorganisms and other biological materials, *CRC Crit. Rev. Microbiol.*, p. 109, 1978.
215. *Problémes d'Ionization et d'Aéro-Ionization*, Rager, G. R., Ed., Maloine S. A., Editeur, Paris, 1975.
216. *Aeroionotherapy*, Gualtierotti, R., Kornblueh, I. H., and Sirtori, C., Eds., Carlo Erba Foundation, Milan 1968.
217. **Wehner, A. P.**, Electro-aerosol therapy, *Am. J. Phys. Med.*, 41, 24, 1962.
218. **Wehner, A. P.**, What is electro-aerosol therapy? *Inhal. Ther.*, 8, 18, 1963.
219. **Himstedt, R.**, Über die Beeinflussung des vegetativen Nervensystems durch die elektrische Aerosolbehandlung, *Arch. Phys. Ther.*, (Leipzig), 3, 198, 1953.
220. **Schulz, K. H.**, Die Bedeutung der Luftelektrizität, der Elektro-Aerosoltherapie, der direkten elektrostatischen Aufladungstherapie nach Takata und ihre biologische Wirkung auf das vegetative Nervensystem unter Berücksichtigung des Vegetonogramms, *Z. Biol. Aerosol-Forsch.*, 12, 206, 1965, Schulz, K. H., *Z. Biol. Aerosol-Forsch.*, 12, 455, 1965.
221. **Bisa, K.**, Die Reagibilität des menschlichen Organismus auf unipolar aufgeladene Aerosole, dargestellt am Verhalten der Flimmer-Verschmelzungsfrequenz, *Z. Aerosol-Forsch.*, 3, 39, 1954.
222. **von Wasielewski, E., Gillissen, G., Himstedt, R., and Genth, W.**, Die biologische Bedeutung der Elektroaerosole unter besonderer Berucksichtigung des HVL-NNR-Systems, *Z. Aerosol-Forsch.*, 4, 18, 1955.
223. **Wehner, A. P., Sehmel, G. A., Schwendiman, L. C., and Harter, J. G.**, Unipolarly charged aerosols. Effect on deposition in tubes and on droplet size, *Arch. Environ. Health*, 20, 28, 1970.
224. **Loeb, L. B.**, *Fundamental Processes of Electrical Discharge in Gas*, John Wiley & Sons, New York, 1939.
225. **Kingdon, K. H.**, Possible biological effects of electrically charged particles in tobacco smoke, *Nature (London)*, 189, 189, 1961.
226. **Bisa, K. and Weidemann, J.**, Die Elektro-Aerosole, *Z. Aerosol-Forsch.*, 3, 220, 1955.
227. **Thews, W.**, Erfahrungen mit der Elektroaerosoltherapie bei Verwendung von Wiesbadener Kochbrunnenwasser, in *Bekämpfung der Silikose*, Lange, F., Ed., Verlag Gluckauf, Essen, 1956.
228. **Cauer, H. and Neymann, N.**, Die Inhalieranlage nach Barthel-Küster auf der Zeche Hannibal und die Inhalierung von Calziumsole gegen Silikosebeschwerden, *Glückauf!* (Essen), 87, 1011, 1951.
229. **Wehner, A. P.**, Die Entwicklung der Elektroaerosologie in den USA, *Z. Biol. Aerosol-Forsch.*, 13, 1, 1966.

230. **Ellis, G. R., Swanson, A. A., and Wehner, A. P.,** Preliminary evaluation of electro-aerosols in the treatment of pulmonary disorders, *Arch. Phys. Med.,* 45, 364, 1964.
231. **Ellis, G. R.,** Preliminary evaluation of electro-aerosols in the treatment of chronic bronchitis, *Arch. Phys. Med.,* 46, 767, 1965.
232. **Barron, D. I. and Dreher, J. J.,** Effects of electric fields and negative ion concentrations on test pilots, *Aerosp. Med.,* 35, 20, 1964.
233. **Chiles, W. D., Cleveland, J. M., and Fox, R. E.,** Study of the effects of ionized air on behavior, Tech. Rep. WADD 60-598, Wright-Patterson Air Force Base, Ohio, 1960.
234. **Chiles, W. D., Fox, R. E., Rush, J. H., and Stilson, D. W.,** Effects of ionized air on decision making and vigilance performance, Tech. Rep. MRL-TDR-62-51, Wright-Patterson Air Force Base, Ohio, 1962.
235. **Zylberberg, B. and Loveless, M. H.,** Preliminary experiments with ionized air in asthma, *J. Allerg.,* 31, 370, 1960.
236. **Blumstein, G. I., Spiegelmann, J., and Kimbel, P.,** Atmospheric ionization in allergic respiratory disease, *Arch. Environ. Health (Chicago),* 8, 818, 1964.
237. **Lefcoe, N.,** Ventilatory function after exposure to ionized air, *Arch. Environ. Health (Chicago),* 7, 664, 1963.
238. **Turiel, I.,** *Indoor Air Quality and Human Health,* Stanford University Press, Stanford, Calif., 1985.
239. **Bossert, W. H. and Wilson, E. O.,** The analysis of olfactory communication among animals, *J. Theor. Biol.,* 5, 443, 1963.
240. **Wadden, R. A. and Scheff, P. A.,** *Indoor Air Pollution,* John Wiley & Sons, New York, 1983.
241. **Walsh, P. J., Dudney, C. S., and Copenhaver, E. D.,** *Indoor Air Quality,* CRC Press, Boca Raton, Fla., 1984.
242. **Gammage, R. B. and Kaye, S. V.,** *Indoor Air and Human Health,* Lewis Publishers, Inc., Chelsea, MI, 1985.
243. **Meyer, B.,** *Indoor Air Quality,* Addison-Wesley Publishing Co., Inc., Reading, MA, 1983.

INDEX

A

H

I

K

L

M

N

O

P

R

S

T

U

V

W

X

Z